STRATEGIES for PHYSICAL MAPPING

Computer Analysis of *Escherichia coli* Restriction Maps

Kenneth E. Rudd,[1] Gerard Bouffard,[2] and Webb Miller[3]

[1]National Center for Biotechnology Information
National Library of Medicine
National Institutes of Health
Bethesda, Maryland 20894

[2]Department of Microbiology and Immunology
George Washington University
Washington, DC 20037

[3]Department of Computer Science
The Pennsylvania State University
University Park, Pennsylvania 16802

A restriction map of the *E. coli* genome has been produced using partial restriction digests of 3400 bacteriophage λ clones generated with eight restriction enzymes (Kohara et al. 1987). Shortly after publication of the Kohara map, one of us (K.E.R.) began the process of pinning all available *E. coli* DNA sequence to the genomic restriction map. The first integrated collection of nonoverlapping sequences and their genomic map positions was released in early 1990.

The project required creation of novel computer software to digitize, process, analyze, and display *E. coli* restriction map data (Miller et al. 1990, 1991; Rudd et al. 1990, 1991). There is little current agreement on how restriction maps should be represented or displayed. We have shown five possible formats for representing a fragment of DNA and some of the restriction sites it contains; these formats are used by the map-handling software discussed herein.

This software and the integrated *E. coli* map allow us to inform the *E. coli* research community of precisely which regions have been sequenced, to automatically position and orient many new sequences on the genomic map, to produce publication-quality figures of all or a portion of the map, and to discover interesting facts about the *E. coli* genome. We believe that the methods described here will serve as the basis for more inclusive compilations of *E. coli* data and for analogous data sets from other organisms.

Specific topics include:

❑ computer analysis and graphical display of restriction maps

❑ integration of the *E. coli* genomic restriction map with DNA sequences

❑ organization of the integrated map data set into database management systems

❑ frequency distribution of genomic restriction sites

❑ genomic map positions of infrequently occurring restriction sites

❑ comparison of the *E. coli* genomic restriction map with the genetic map

INTRODUCTION

A restriction map of the *E. coli* genome has been produced using partial restriction digests of 3400 bacteriophage λ clones generated with the eight restriction enzymes *Bam*HI, *Hind*III, *Eco*RI, *Eco*RV, *Bgl*I, *Kpn*I, *Pst*I, and *Pvu*II (Kohara et al. 1987). The restriction sites in this map are of sufficient density and accuracy relative to the genome's size that the map position of a DNA sequence containing one or two genes can often be determined automatically. This approach is the basis of our ongoing effort to develop new computer methods to digitize, process, analyze, and display *E. coli* restriction map data (Miller et al. 1990, 1991; Rudd et al. 1990, 1991). The current versions of the data sets we use to represent the *E. coli* genome (EcoSeq, EcoMap, EcoGene) reflect the fact that 38% of the *E. coli* genome is DNA sequenced (as of December 1, 1991). We have integrated these data sets using software designed specifically for this purpose, resulting in a revised, integrated genomic restriction map designated Ecoli5.map (Rudd 1992), which is graphically displayed in a highly abbreviated form in Figure 1.

To construct this integrated map, we analyze published restriction maps derived using gel electrophoresis methodology as well as restriction maps predicted from DNA sequence. These maps are either limited to a particular region of the *E. coli* chromosome (local restriction maps) or they encompass the entire genome (genomic restriction maps). Finally, we distinguish between low-resolution maps derived by using rare cutters (restriction enzymes that cut infrequently, e.g., *Avr*II, see below) and high-resolution restriction maps made with enzymes that cut fre-

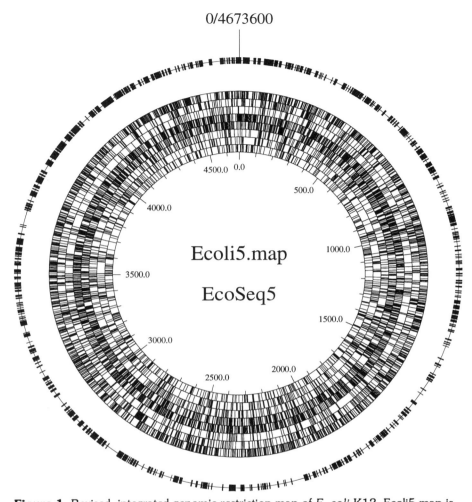

Figure 1 Revised, integrated genomic restriction map of *E. coli* K12. Ecoli5.map is depicted with the sequenced regions (38%) indicated by blackened areas on the outer circle. The order of restriction enzyme sites (from outer to inner circles) is *Bam*HI, *Hind*III, *Eco*RI, *Eco*RV, *Bgl*I, *Kpn*I, *Pst*I, *Pvu*II.

quently, such as the *E. coli* genomic restriction map that is at the center of this analysis.

Development of computer methods to handle restriction maps lags far behind analogous methods for sequence data. Whereas most published DNA sequences can be readily obtained from a database, restriction enzyme maps must be digitized from published journal figures or obtained directly from authors as computer files. Such an electronic form of DNA sequence and restriction map data is required for computer storage and analysis, electronic communication, automatic conversion to alternative forms, and precise display. Unfortunately, there is little current agreement on how restriction maps should be represented or displayed. In Figure 2 are shown five possible formats for representing a fragment of DNA and some of the restriction sites that it contains; these formats are used by the map handling software discussed below.

OUR INTEGRATED GENOMIC DATA SET

The *E. coli* genomic restriction map and miniset clones

The Kohara genomic restriction map and clones of the *E. coli* chromosome are of immense value to *E. coli* geneticists. An ordered "miniset" of 476 overlapping λ clones is widely distributed. The commercial availability of DNA filters containing the miniset clones has already supplemented traditional genetic crossing methods used to locate new genes by enabling the use of DNA hybridization methods. In addition, Kohara et al. (1987) aligned published restriction maps of *E. coli* clones and sequence-derived maps to their consensus genomic map, allowing the genes encoded by these DNA fragments to be positioned. The integration of DNA sequence and restriction map information with genetic map data was pioneered by Barbara Bachmann as she considered all the available data in the construction of *E. coli* genetic maps (Bach-

Figure 2 Formats for representing and displaying restriction maps. The EMBL DNA sequence entry ECNARZYW (X17110) was used to generate a restriction map which was then aligned and integrated into the genomic restriction map. (*A*) Textual representation consisting of the positions (in bp) and cognate enzyme names (e.g., *Eco*RI) of a set of restriction enzyme recognition sites, preceded by a sequence length (in bp). The position of a restriction site refers to the base at the site's 5 ′ end. A local map in this format (which we call a "Probe") is used by our MapSearch program (see text) to search a genomic restriction map (also in this format) for matches. (*B*) A sequence of one-letter codes for restriction enzymes, with site addresses not indicated. This incomplete representation was of interim value for aligning restriction maps (Rudd et al. 1990), since it permitted use of amino acid alignment programs such as FASTA (Pearson and Lipman 1988) and BLASTP (Altschul et al. 1990). (*C*) The DNA sequence format, using the ambiguous DNA base symbol "N." These maps are useful as surrogates for unsequenced or confidential regions of genomic DNA when creating large melds of DNA sequence (see the discussion of the Mask and BigSeq programs). A portion of ECNARZYW derived sequence map is shown. The "one-line" (*D*) and "eight-line" (*E*) graphic display formats used by the PrintMap program (see text). The sites above the ECNARZYW feature span are all derived from the DNA sequence. Flanking sites are from the genomic restriction map. The *narZYWV* genes are also depicted as feature spans, including the direction of translation. The genomic kilobase scale is shown above the restriction sites in *E* and below them in *D*. A physical scale in units we define as centisomes (1% of the chromosome length) is shown below the restriction sites in *E*. Miniset clones are also depicted as feature spans. CloneConvert and MapSearch were used to define the orientation (+ or –) of the chromosomal inserts in the λ cloning vector (see text); clones in a + orientation have the left arm of the λ vector to the left of the insert as depicted.

A

7080

```
 522 BglI
 746 PvuII
2222 PvuII
2258 BglI
2517 PstI
2591 EcoRV
3851 PvuII
4548 PvuII
4565 BglI
5811 PstI
5814 PvuII
6248 EcoRV
6289 PvuII
6596 EcoRV
6870 EcoRV
```

B   GVVGSFVVGSVFVFF

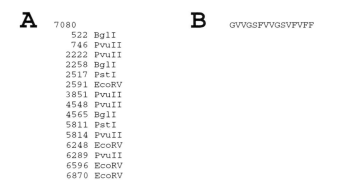

C

```
NNNNNNNNNNNNNNNNNNNNNNNNNNNNNNNNNNNNNNNNNNNNNNCAGCTGNNNNNNNNN
NNNNNNNNNNNNNNNNNNNNNNNGCCNNNNNGGCNNNNNNNNNNNNNNNNNNNNNNNNNNN
NNNNNNNNNNNNNNNNNNNNNNNNNNNNNNNNNNNNNNNNNNNNNNNNNNNNNNNNNNNNN
NNNNNNNNNNNNNNNNNNNNNNNNNNNNNNNNNNNNNNNNNNNNNNNNNNNNNNNNNNNNN
NNNNNNNNNNNNNNNNNNNNNNNNNNNNNNNNNNNNNNNNNNNNNNNNNNNNNNNNNNNNN
NNNNNNNNNNNNNNNNNNNNNNNNNNNNNNNNNNNNNNNNCTGCAGNNNNNNNNNNNNNNN
NNNNNNNNNNNNNNNNNNNNNNNNNNNNNNNNNNNNNNNNNNNNNNNNNNNNNNNNNGATAT
CNNNNNNNNNNNNNNNNNNNNNNNNNNNNNNNNNNNNNNNNNNNNNNNNNNNNNNNNNNNNN
```

D

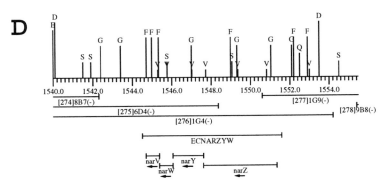

E

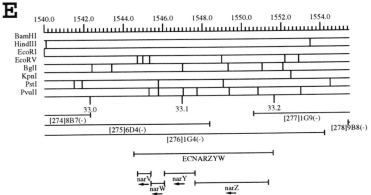

Figure 2 *(See facing page for legend.)*

mann 1983, 1990). The Kohara map made possible a comprehensive, computer-based integration of these data.

To begin this integration process, we digitized the Kohara map (Rudd et al. 1990), as had other groups with similar goals (Medigue et al. 1990b; Watanabe and Kunisawa 1990; Churchill et al. 1990). Our digitization procedure was performed on an enlarged copy of the printed Kohara map that he provided to us. The Kohara map depicts sites for each of the eight enzymes on a separate line (as in Fig. 2E). These eight different maps were digitized separately and then combined into one genomic map file, designated Ecoli.map. We also digitized the endpoints of the miniset clones as depicted on the original map. Dr. Katsumi Isono kindly provided us with a computer file of the individual restriction maps associated with the miniset clones. We processed this file into MapSearch Probes (digital restriction maps, see Fig. 2A), which were positioned and oriented on the genomic map to give an alternative representation of the miniset clones; see the discussions of MapSearch and CloneConvert, below.

Although we resolved the restriction sites depicted on the Kohara map figure to multiples of 100 base pairs, the map itself is not this accurate. The range of fragment sizes measured in Kohara's partial digestion method is 10–30 kb due to the addition of the 9-kb right arm of the λ vector. The difficulty in measuring fragments of this size and other factors concerning the overall accuracy of the Kohara map have been discussed previously (Churchill et al. 1990; Daniels 1990b; Medigue et al. 1990a,b; Rudd et al. 1990). Some regions of the Kohara map are considerably less accurate than the bulk of the map, possibly due to mapping and cloning artifacts resulting in gaps and local distortions (Kohara et al. 1987; Medigue et al. 1990a,b; Rudd et al. 1990, 1991; Rudd 1992). The filling of these gaps and the correction of distorted regions is progressing rapidly due to the availability of alternative *E. coli* genomic maps (Daniels and Blattner 1987; Knott et al. 1988, 1989; Birkenbihl and Vielmetter 1989b), local restriction map corrections (e.g., Lipinska et al. 1989; Oh et al. 1990; Conway et al. 1991), and the increasing availability of DNA sequences for these regions (Kroger et al. 1991; Medigue et al. 1991; Rudd 1992). Forty regions of the original Kohara map were cited as containing no information about the presence or absence of *Eco*RV sites. These regions are indicated with dotted lines on the original Kohara map and were digitized and reported by Rudd (1992). Despite all this, we consider the Kohara map to be a remarkable achievement that represents most regions of the *E. coli* chromosome quite accurately.

The revised genomic restriction map

The original Kohara restriction map was derived from a particular strain of *E. coli*, W3110. This strain differs from many other *E. coli* strains, including MG1655, which is a derivative of the original *E. coli* K12 isolate, EMG2. MG1655 lacks the F fertility plasmid and the λ prophage that were present in the original K12 isolate, but it should otherwise be nearly identical to the wild type. MG1655 is being used in the *E. coli* genome sequencing project spearheaded by Fred Blattner and Donna Daniels.

We revised the original Kohara map to more closely represent MG1655, to record some published corrections of certain regions, and to fill gaps present in the original Kohara map. In particular, the large IN(*rrnD-rrnE)1* inversion present in W3110 and the Kohara map was reverted (Rudd et al. 1990), the *htrA* locus was corrected, and four duplicate copies of the *tdc* operon and *rnpB* gene were removed. We refer the reader to other articles where we have discussed these revisions in detail (Rudd et al. 1990, 1991; Rudd 1992). A recently developed program, GeneScape (Bouffard et al. 1992) (see below), facilitates rearrangements of the genomic restriction map and its associated features, thereby allowing individual researchers to easily create customized versions of the *E. coli* genomic map.

The integrated genomic map

We use the program AlterMap to replace sections of the revised genomic map with maps derived from DNA sequence, as discussed in more detail in the AlterMap section. The resulting integrated map has clones and DNA sequences aligned to it (called "feature spans") whose endpoint positions must be revised whenever the integration of a DNA segment alters the map's length. Only feature spans between the integration site and the end of the map need to be altered. (Although the *E. coli* chromosome is circular and has no real end, we use a linear representation, which does have an endpoint that is identical in vivo to its start point). DNA sequences used in this integration process are derived from a variety of different *E. coli* K12 strains (and in a very few cases *E. coli* B). Thus, the revised map no longer represents just one strain, W3110, and differences between the sequence-generated segments and the original Kohara map may reflect DNA sequence polymorphisms and strain-specific rearrangements (e.g., insertion elements and inversions) as discussed above. We still align the miniset clones, derived from W3110, which can now be thought of as containing DNA highly homologous to sequenced regions. Although we continue to make the original digital Kohara map available for those who desire it, we believe that the integrated map should be used in most cases. Integrating DNA sequences into the genomic restriction map allows us to accurately position the gene-coding regions and other sequence features on the genomic map. This has been done in a project we refer to as EcoGene, and the resulting genomic positions are part of the EcoGene data set. Several new programs were developed for this purpose, particularly one called Global that converts local DNA features to genomic sequence features. The EcoGene project will be described in more detail elsewhere (K. Rudd and W. Miller, in prep.).

Other *E. coli* genomic maps

We emphasize that the integrated genomic map with clones, DNA sequences, and genes aligned to it is just one of several possible views of the *E. coli* chromosome. Other views, such as the original Kohara map, revisions of the Kohara map using another computerized approach (Medigue et al. 1990a,b, 1991), independently derived high-resolution *E. coli* genomic restriction maps (Daniels and Blattner 1987; Knott et al. 1988, 1989; Birkenbihl and Vielmetter 1989b; Daniels 1990b), and the *E. coli* genetic map (Bachmann 1990), provide alternatives that are very useful in identifying possible strain polymorphisms or errors in the existing data sets. These discrepancies include differences in map position, local gene order, and gene orientation. In some cases, they may be due to alternative interpretations of published data; in other cases, strain differences or instabilities may be the cause. Some differences between the Kohara map (K12 strain W3110) and chromosomal restriction maps of other strains have been discussed previously (Daniels and Blattner 1987; Knott et al. 1988, 1989; Birkenbihl and Vielmetter 1989b; Daniels 1990b). We hope that efforts to refine these alternative maps continue in parallel with our integrated map development. We also hope that other groups involved with genomic map revision begin to exchange data sets freely so that the public can benefit from the accelerated genomic map refinement that would likely result from this increased communication. In a later section, we compare our placement of genes on the integrated genomic map with the 1990 genetic map on a global scale, an extension of our earlier comparative analysis (Rudd et al. 1990).

COMPUTER SOFTWARE FOR MANIPULATING RESTRICTION MAPS

We and our collaborators have developed a number of programs to digitize, display, manipulate, and analyze *E. coli* restriction maps. Figure 3 depicts the relationships among some of these programs in a data flow diagram. These programs are freely available upon request or by computer from the Internet anonymous ftp site ncbi.nlm.nih.gov in the Eco/EcoMap/Code subdirectory. We encourage the reader to experiment with the Macintosh versions of MapSearch, PrintMap, ProbeMaker, and GeneScape (see below). In what follows, we describe some of our programs in more detail, since they are the concrete embodiment of our approach to computer analysis of restriction maps.

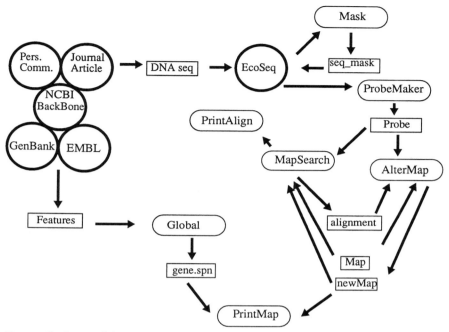

Figure 3 Some of the programs (ovals), computer files (rectangles), and DNA sequence databases and sources (circles) used in our approach to aligning and integrating DNA sequences to the *E. coli* genomic restriction map. DNA sequences obtained from a variety of sources are collected into the EcoSeq data set. (Several programs process these sequences, e.g., to remove any remaining vector sequence and to eliminate overlaps, but these programs are not represented here.) The program Mask protects confidential sequences. ProbeMaker computes the restriction map for a given DNA sequence. MapSearch searches the genomic restriction map for regions that approximately match a local restriction map, and the resulting Probe-to-Map alignments can be viewed with PrintAlign (Rudd et al. 1991). AlterMap replaces a portion of the genomic map with the local map computed from a DNA sequence. Global converts addresses of genes within a sequence to positions on the genomic map. Maps such as Fig. 2, D and E, are prepared by the PrintMap program. The NCBI BackBone database is produced as a joint effort between the Library Operations division of the National Library of Medicine and the NCBI; it contains DNA sequences entered directly from publications.

ProbeMaker

ProbeMaker converts DNA sequence files in FASTA format to digital restriction maps used as MapSearch Probes (see below). It searches for the eight enzymes in the genomic restriction map by default, but takes an optional input file, enz.lst, that can contain any set of restriction site specificities. For instance, we use ProbeMaker to identify all known restriction sites in a target DNA sequence using a very large enz.lst file, or to determine the rare restriction enzyme recognition sites in EcoSeq (see below). ProbeMaker was designed to recognize any of the IUPAC/IUB ambiguity codes in a target DNA sequence. For example, the ambiguity code D in GDGCHC, the *Nsp*II recognition site, represents A, G, or T (not C). In a target DNA sequence, these bases could be represented as D, A, G, T, R (A or G), W (A or T), or K (G or T). ProbeMaker locates regions of an ambiguous DNA sequence that are guaranteed to be restriction sites; possible but uncertain sites are not reported.

MapSearch

MapSearch (Miller et al. 1990, 1991; Rudd et al. 1990, 1991) locates the regions of a genomic restriction map that most closely resemble a given local restriction map, which we call a "Probe." The Probe is usually derived from a DNA sequence using the program ProbeMaker. Map-Search scans the genomic "Map" file for matches to a Probe in both orientations and returns a list of possible alignments ranked by close-ness of fit. Alignments can also be displayed, indicating which specific Probe and Map sites are paired (Fig. 4). If sequenced genes are part of a large (5 kb or more) contig of DNA sequence, the alignment process is quite reliable.

MapSearch optionally aligns the Probe to a user-specified number of randomly shuffled Maps, and uses extreme-value distribution theory (Gumbel 1962; Rudd et al. 1990) to compute the probability of an align-ment's score occurring by chance (see the "prob" column of Fig. 4A). We usually calculate p values using 100 Map shuffles, although this in-creases computation time 101-fold. Reasonable values can be obtained using 10 Map shuffles. The "st dev" column in Figure 4A refers to all pos-sible alignments of the Probe to various regions of the Map; although it is less useful than the p value, its computation does not involve shuffling the Map.

Any number of MapSearch alignments can be displayed, which is useful when looking for the locations of small repeated DNA sequences, like the *E. coli* insertion elements. More typically, only one alignment is biologically correct. In questionable cases, the correct alignment must be determined using additional laboratory data, such as the genetic map position, hybridization to the Kohara miniset, or a gel-derived Map-Search Probe of the DNA surrounding the sequenced region.

Probes being aligned to the Kohara map must use a subset of the eight Kohara enzymes. MapSearch can be instructed to read a file of relevant enzyme names, although the default is to use only enzymes whose names appear in the Probe file. This is useful for Probes not derived from DNA sequence. However, if a DNA fragment was treated with a restriction enzyme and found NOT to contain a site for one of the enzymes named in the Map file, this enzyme should be in the enzyme-

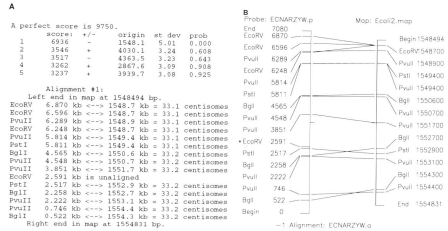

Figure 4 (A) Sample MapSearch output using a Probe generated from the *narZYWV* operon DNA sequence (EMBL X17110). The user has asked that MapSearch display only the highest-scoring alignment, together with summary statistics (score, orientation, position, and two reliability estimates) for the top five. In the highest-scoring alignment, one Probe site is unaligned (indicated by an asterisk), three *Eco*RV sites in the Probe are aligned to the same Map site, and three aligned pairs "crisscross" (e.g., the *Pvu*II site at 746 bp and the *Bgl*I at 522 bp are aligned to Map sites that lie in the reverse order). Genomic sites are reported in both kilobases and centisomes (*see* above). See the text for an explanation of the "st dev" and "prob" statistics. (B) Graphical interpretation of the alignment in part A rendered by the PrintAlign program.

name file (along with each enzyme type contained in the Probe). In the construction of the Kohara map, *Eco*RV cut the *E. coli* chromosome with only 50% efficiency, so we often rerun unsuccessful MapSearch alignments using an enzyme-name file that lacks *Eco*RV.

We expended considerable effort investigating appropriate ways of defining and scoring alignments between a Probe and a Map. The first attempt ignored the distance between successive sites (Fig. 2B). Later we tried several variations of a dynamic-programming algorithm of Michael Waterman (Waterman et al. 1984). Due primarily to the need for the alignment of multiple Probe sites to singular Map sites and the ability to crisscross aligned pairs, we eventually developed a more successful algorithm, called MapSearch, for positioning short sequence-derived Probes on a gel-derived Map. We have described the details and evolution of our algorithm previously (Miller et al. 1990, 1991; Rudd et al. 1990, 1991).

One parameter used by MapSearch, called BETA, sets the penalty, measured in base pairs, assigned to each unaligned site. This provides a means of limiting address discrepancies found in optimal alignments, since the penalty for aligning two distant sites can be greater than the sum of the two unaligned-site penalties (or one, if the Map site is already aligned to another Probe site) that result if the sites are left unaligned.

The default value of 650 for BETA was empirically determined to be suitable for short sequence-derived Probes (Miller et al. 1990). BETA values of 1000–2000 should be used for large (in excess of 20 kb) Probes.

We have overhauled the MapSearch C-language code to conform to recently developed standards (NCBI CoreTools) designed to achieve consistent cross-platform compatibility; this permits the incorporation of a portable graphic user interface (J. Ostell and J. Kans, unpubl.). Planned work includes development of a version of MapSearch geared for longer Probes and for gel-derived Probes. A recent algorithm of Huang and Waterman (1992) is a candidate for such applications.

CloneConvert

As mentioned above, we were provided with a data file containing individual restriction maps for the Kohara miniset. This file, called miniset.dat, contains 476 of the 3400 clone restriction maps used to formulate the Kohara map. We wrote a program, CloneConvert, that converts a file in the miniset.dat format to individual MapSearch Probes, ignoring the restriction sites in the arms of the bacteriophage λ vectors. We have used these alignments to determine orientations for the clones on the genomic restriction map (Fig. 2D,E). Positive orientations (+) of λ clones are defined as having the left arm of λ on the clockwise side of the insert as depicted on the integrated map. The positions obtained with MapSearch differ from the endpoints digitized from the original figure by as much as 5 kb (K.E. Rudd, unpubl.). They are helpful for resolving discrepancies concerning containment of a gene or restriction site in a clone. This is important, since a major use of the Kohara map is to identify the miniset clones that contain certain genes.

AlterMap and AlignMaker

AlterMap uses MapSearch alignments to replace sections of the Kohara map with local maps, which are typically derived from DNA sequence (Rudd et al. 1991; Rudd 1992) and hence are generally more accurate than the Kohara map for reasons discussed elsewhere in this paper. (In a later section, we estimate the efficiency with which the eight Kohara enzyme sites were detected in the original map.) The command line for AlterMap is as follows:

AlterMap MOVE=n k Alignment Probe Map [SpanFiles]

MOVE is a recently added AlterMap option that helps prevent the inappropriate removal of unaligned Map restriction sites that have been slightly misplaced so that they fall within the edges of the Probe's "shadow" (i.e., the region of the Map that corresponds to the Probe under the chosen alignment). MOVE bumps unaligned sites lying within n bases of an end of the shadow to a position just outside the shadow and alerts the user, who may choose to discard the site. We typically set MOVE=500. The k parameter specifies which MapSearch alignment to use for the Probe integration (usually k=1 for Probes that exceed 3 kb in length). The Alignment is usually a MapSearch output file, but the output of our new AlignMaker (see below) program can also be used. The specified Map must be the exact version of the genomic map used to generate the MapSearch or AlignMaker Alignment. Thus, MapSearch and AlterMap are used in an alternating fashion during the integration process. The SpanFiles describe various Map feature intervals, such as the miniset clones, regions without *Eco*RV site information, and DNA sequences, which must be simultaneously updated. These same SpanFiles are also used as GeneScape and PrintMap input (see below).

In some cases, an appropriate MapSearch alignment cannot be easily obtained. Perhaps a DNA sequence contains too few restriction sites, or the genomic map corresponds poorly to the sequence-derived map due to errors or strain polymorphisms. A new program, AlignMaker, produces a file resembling MapSearch output that can be read by AlterMap, but where the local map's genomic position is specified by the user. This allowed us to integrate 110 DNA sequences at positions determined by methods other than MapSearch alignment, such as mapping the tRNA genes to the Kohara miniset by hybridization (Komine et al. 1990).

Mask and BigSeq

Mask and BigSeq are new programs that produce ambiguous DNA sequence files, resembling Figure 2C. Mask, written by our collaborator Marvin Shapiro, protects confidential, unpublished DNA sequences in EcoSeq. It replaces the bases of a confidential sequence, or region of a sequence meld, with the ambiguous DNA base "N," except for recognition sites of the eight enzymes used in the Kohara map. This incorporates unpublished information into the EcoSeq and EcoMap updates, informing the *E. coli* research community of precisely what regions of the chromosome have been sequenced. Interested parties can then contact the authors of the unpublished data. We hope that this approach will reduce unnecessary duplication of effort and encourage prepublication collaborations.

As completion of the *E. coli* genome sequencing project approaches, we are developing a new strategy for integrating genomic restriction map and DNA sequence data to augment the MapSearch/AlterMap approach outlined above. With 38% of the genome already sequenced, the likelihood that new DNA sequence will overlap with well-positioned genomic DNA sequence increases both statistically and by design (targeted gap-filling). Our collaborator Carolyn Tolstoshev has written a program called BigSeq that creates a file whose current version, called Ecoli5.seq, contains 4,673,600 bp in a single contig. Input for BigSeq includes the integrated genomic map, the entries in the EcoSeq data set, and the genomic map positions of the EcoSeq entries. Although most of the base pairs in Ecoli5.seq are "N"s, this representation anticipates completion of the *E. coli* genome sequence. As with Mask, unsequenced portions include the recognition sequences for all the Kohara enzyme sites (Fig. 2C). ProbeMaker converts the Ecoli5.seq file into the corresponding genomic map file, Ecoli5.map, so we can now combine the DNA sequence melding procedure (Rudd et al. 1991; Rudd 1992) and the updating of the integrated genomic map into a single procedure. As new sequences are melded into Ecoli5.seq, the feature spans associated with the genomic sequence are simultaneously updated. The new integrated map is then generated using ProbeMaker. While we are developing new software to perform this task, we are also experimenting with use of a commercial Macintosh program (Gene Construction Kit; Textco, Inc.) designed to update and track changes to DNA plasmid cloning constructs.

PrintMap

PrintMap is a program developed by Craig Werner, based on his Plasmid Description Language (PDL), to produce high-quality restriction map depictions in PostScript code (Rudd et al. 1991). (PostScript is a computer language developed by Adobe Systems Incorporated that is utilized by many laser printers.) The PrintMap command line takes the form:

PrintMap -b startbp -f endbp -i Map_file -o PDL_outfile -Z StyleSheet

Any segment (from startbp to endbp) of the input Map_file, together with specified files of feature spans (e.g., clone, sequence, or gene endpoints), is processed to a PDL output file, which can be sent to a laser printer or a PostScript screen display program (such as the Unix "postscript," "xpsview," or "pageview" utilities) or stored as a PostScript text file. The StyleSheet contains instructions that control many parameters of PrintMap output, including the number of map lines per page, point size, margins, label spacings, tic distance, line spacings, line width, page orientation (Landscape or Portrait), and format (1-line like Fig. 2D, or 8-line like Fig. 2E). Almost any map style desired can be produced using an appropriate StyleSheet, which can be saved for future use. PrintMap was used to prepare Figure 2, D and E, as well as a 16-page map of the entire genome (Rudd 1992). Circular genomic maps can be depicted using a new capability, which we have employed for Figures 1 and 8, as well as sequenced mitochondrial and chloroplast genomes, and viral genomes such as T4, λ, and sequenced eukaryotic viruses (data not shown).

MapShow

We currently run the tools described above (and others) under the Unix operating system, which makes it straightforward to package commonly used combinations of these commands as a new command. For example, MapShow is the following simple Unix shell script that we use to dynamically create MapSearch alignments and graphically render them with PrintAlign, which draws Probe-to-Map alignments in a Tektronics 4014 graphics window on a Sun workstation.

```
(move a copy of a sequence file to the current directory)
cp ECOSEQ/$1.seq .;
(create a Probe)
ProbeMaker $1.seq;
(run MapSearch using the revised, not integrated map)
MapSearch -e all.enz -p $1.p -m Ecoli2.map -o $1.o -p;
(convert MapSearch output to PrintAlign input with the M2P program)
M2P 1 <$1.o>$1.a;
(run PrintAlign)
PrintAlign -f +1 -d -m Ecoli2.map -p $1.p -a $1.a > $1.qwd;
(run the Unix qdraw utility program)
qdraw -p tk $1.qwd;
(clean up the temporary files)
rm $1.p $1.a $1.o $1.seq $1.qwd;
```

MapShow is invoked with a command like "MapShow ECNARZYW," which displays Figure 4B on the computer screen. A variation of Map-Show allows the user to view discrepancies between an individual mini-set clone map in Isono's miniset.dat file (see above) and either the Kohara consensus map or the integrated genomic map. Another variation allows the user to designate any interval (up to about 50 kb in length to be readable) of any map in the single-contig DNA sequence format (like Ecoli5.seq) to be aligned to any digital genomic restriction map (like Ecoli2.map).

E. COLI RESTRICTION MAP DATABASES

GeneScape

GeneScape (Bouffard et al. 1992) is a genomic map database program for the Macintosh, developed using the FoxBase+/Mac database development system. It has three basic functions: (1) a restriction site editor (Fig. 5); (2) a search function that quickly locates miniset clones, DNA sequences, and genes aligned to the genomic restriction map; and (3) a versatile on-screen graphic map display function (Fig. 6) that allows the user to zoom from a display of the entire genome down to base-pair resolution. Graphical maps can easily be sent to laser or dot-matrix printers for documentation or publication. Maps and SpanFiles are provided as ASCII text files which can be imported into the database. Facilities are provided to allow users to create their own Maps, which can then be exported for use in MapSearch or PrintMap. Users are provided with regular updates of the *E. coli* integrated Map and associated SpanFiles of clones, sequences, and genes, preferably via electronic mail or by anonymous ftp protocols. Floppy diskette distribution is also available. Registered users become part of an *E. coli* information exchange network, designated ColiNet. GeneScape can be used to display and organize genome data from any organism that has a genomic restriction map and/or significant amounts of genomic DNA sequence.

 File   Edit   Maps & Probes   Utilities

Restriction Site Set

Set ID: 7   Length: 48502
 Comment: Prophage form of Lambda
 This set is not an original submission.

<-- Previous Set | Go to Set | Next Set -->

	Serial	Address	Enzyme	Prev Set	Prev Ser	Prev Add	
TOP	51	16412	HindIII	4	200	44141	**COPY**
	52	16904	HaeII	4	201	44633	
//	53	17243	EcoRI	4	202	44972	**CUT**
	54	17950	HaeII	4	203	45679	
/	55	18070	HaeII	4	204	45799	**ADD**
	56	18097	EcoRV	4	205	45826	
GOTO	57	18148	HaeII	4	206	45877	**ERASE**
	58	18709	ClaI	4	207	46438	
	59	20982	PvuII	4	1	209	
\/	60	21177	BglI	4	2	404	**PASTE**
	61	21188	BglII	4	3	415	
\\//	62	21231	MluI	4	4	458	
	63	21423	EcoRV	4	5	650	
	64	21629	HaeII	4	6	856	
BOTTOM	65	21873	HaeII	4	7	1100	
	66	22641	HaeII	4	8	1868	

Figure 5 GeneScape restriction site editor. In preparation for inserting the bacteriophage λ genomic restriction map into the *E. coli* genomic map, so as to model a λ lysogen, the λ map for the vegetative form is converted to that for the prophage form. This involves a "Cut" operation, from bp 1 to 27730 (the center of the λ *attP* attachment site), and a "Paste" operation to swap left and right portions, essentially joining λ at the "sticky ends." For each restriction site, GeneScape shows the index in both the current (prophage form) map and the vegetative-form map.

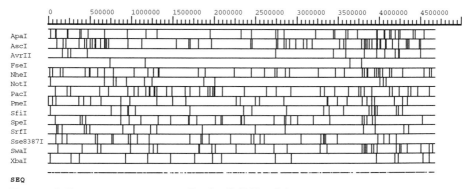

Figure 6 Rare restriction sites in EcoSeq5 (38% of the genome). Darkened regions of the bottom line indicate sequenced portions of the *E. coli* genome. This figure was produced using the GeneScape program (see text).

Prolog

Ross Overbeek has imported the EcoSeq/EcoMap/EcoGene data sets into the Quintus Prolog database management and query system. We use the powerful querying capabilities of Prolog to process and extract data from these data sets, including a suite of integrity checks. The frequency distributions of all nucleotide sequence n-mers (i.e. monomers, dimers, trimers, etc.) of EcoSeq are routinely extracted using Prolog queries, as were the data presented below on the frequency distributions and Markov chain statistics for palindromic hexamers.

Sybase

In collaboration with Jim Ostell and Carolyn Tolstoshev of the National Center for Biotechnology Information, we are developing a data management system using the Sybase relational database system, which utilizes the SQL data query language. Primarily designed as a curator management system, this database will enable us to link to the other Sybase databases of molecular sequence data currently being developed at NCBI. One key aspect of this system is the use of Abstract Syntax Nomenclature (ASN.1) for a rich and structured exchange of data between various databases, such as the GenBank, EMBL, PIR, and Swiss-Prot molecular sequence databases. The NCBI staff are developing many new tools for molecular sequence data processing that we plan to utilize for display, query, and retrieval of *E. coli* sequence data from these other sources as well as from EcoSeq. Our *E. coli* Sybase system will provide links between the restriction map data in EcoMap and molecular sequence data.

ONTOS

Our *E. coli* genomic data sets and programs have recently been organized into an object-oriented database management system, ONTOS, by Dong-Guk Shin and his collaborators (Shin et al. 1992). The nature of our data sets appears to lend itself naturally to the organization of the object-oriented paradigm because of the tight interactions of these complex data sets and the software used to create and manipulate them.

Other *E. coli* databases

A number of other groups are developing management systems for *E. coli* genomic data sets, including DNA and protein sequences (Kunisawa et al. 1990; Danchin et al. 1991; Kroger et al. 1991; Kunisawa and Nakamura 1991), two-dimensional gel data (VanBogelen and Neidhardt 1991), and genetic data (Berlyn and Letovsky 1992; M. Berlyn, pers. comm). We hope to interact with these groups and to foster a climate of open data exchange.

ANALYSIS OF THE INTEGRATED MAP

As discussed above, the EcoSeq5 data set contains 38% of the genomic *E. coli* DNA sequence in nonoverlapping contigs, almost all of which is accurately positioned and oriented on the genomic map. This provides an unprecedented opportunity to obtain accurate answers to a wide range of questions concerning the molecular biology of this model organism. In the remainder of this chapter, we present results obtained from this data set, with emphasis on restriction enzyme cleavage sites. We first investigate the accuracy of the Kohara map and raise a question concerning the distribution of *Eco*RI and *Hin*dIII sites. We then consider the frequency, distribution, and map positions of restriction sites of various rare 6-base and 8-base cutters (especially *Avr*II) and of the 64 palindromic hexamers, 52 of which are recognized by at least one known restriction enzyme (Roberts and Macelis 1991). Finally, we update the comparison of the genomic restriction map with the genetic map and demonstrate how restriction map alignments can locate copies of repeated DNA in *E. coli*.

When interpreting these results, the reader should keep in mind that discrepancies between gel-derived map sites and those derived from EcoSeq5 could arise from a number of causes. With frequent-cutting enzymes, the most likely, as discussed below, is the inability to detect small fragments on gels. In vivo, sites may be methylated or otherwise modified to prevent recognition by restriction enzymes. Moreover, EcoSeq5 is a collection of sequences from different strains of *E. coli*, creating the potential for strain polymorphisms. Finally, in relatively rare instances, sequencing error can either create fictitious sites or mask real ones.

Accuracy of the Kohara map

Sequence-derived restriction maps are more accurate than gel-derived maps, particularly in resolving sites for the same enzyme that lie less than 300–500 bp apart. Using theoretical analyses of the Kohara map's site-distribution patterns, Churchill et al. (1990) estimated that 5–15% of the Kohara map's enzyme sites are in fact multiple sites. Their examination of 213,756 bp of *E. coli* sequence (4.5% of the total genome) generally corroborated this prediction, except for the relatively abundant sites for *Eco*RV, *Pvu*II, and *Pst*I. Samuel Karlin's group has developed new statistical methodologies for analyzing restriction map as well as DNA sequence data (Karlin and Macken 1991a,b; Burge et al. 1992; Karlin et al. 1992). These studies confirm and extend the work of Churchill et al. (1990).

EcoSeq5 provides the largest data set to date for observing the efficiency with which the mapping technique of Kohara et al. (1987) detected chromosomal restriction sites. To estimate how many sequenced restriction sites went undetected in the Kohara map, we compared the numbers of map sites before (Ecoli2.map) and after (Ecoli5.map) integration of sequence-derived sites. (This estimation of the efficiency of detection of closely spaced sites does not take into account that some sites that are not closely spaced to other sites have been added or removed from the map during the AlterMap-facilitated updating procedure.) By determining the exact numbers of sites in the sequenced regions, we calculated the percentage of sequenced sites that were detected by Kohara et al. (1987). For example, consider the row for *Bam*HI in Table 1. There are 473 *Bam*HI sites in Ecoli2.map, 488 in Ecoli5.map, and 194 in EcoSeq5 (aligned subset). Thus, 15 (i.e., 488–473) of the 194 sites were missed, so the efficiency of detection was 179/194, or 92.3%. As shown in Table 1, the procedure of Kohara et al. (1987) detected restriction sites with efficiencies between 85% and 92%, except for *Eco*RV, *Pvu*II, and *Pst*I. This is in accordance with the predictions of Churchill et al. (1990). (The last column of Table 1 is explained below.)

A low efficiency of *Eco*RV site detection is to be expected, since 13% (in 40 regions) of Ecoli2.map contains no information about *Eco*RV sites. Accounting for these regions, plus 15% missed because of close sites, we would expect an efficiency of detection around 74% = (87–[0.15]87)%, as compared to the 54% observed efficiency. This unexpectedly low efficiency may result from the observed effects that nucleotides flanking the recognition pattern have on the cleavage of

Table 1 Counts of Kohara recognition sites

Enzyme	Recog. Site	Ecoli2.map	Ecoli5.map	EcoSeq5[a]	Effic.[b] (%)	Sites Seq'd[b] (%)
*Bam*HI	G^GATCC	473	488	194	92.3	37.8
*Bgl*I	GCCNNNN^NGGC	1559	1667	705	84.7	38.3
*Eco*RI	G^AATTC	601	628	309	91.3	46.9
*Eco*RV	GAT^ATC	1158	1506	789	55.9	38.1
*Hind*III	A^AGCTT	509	545	278	87.1	47.5
*Kpn*I	GGTAC^C	481	502	208	89.9	38.9
*Pst*I	CTGCA^G	833	916	443	81.3	43.2
*Pvu*II	CAG^CTG	1421	1593	753	77.2	40.9

[a]The sites in the few EcoSeq5 entries not yet aligned to the genomic map are omitted.
[b]See text for explanation.

*Eco*RV (and perhaps *Pvu*II) (Medigue et al. 1990a; but see Taylor and Halford 1992). On the other hand, we have observed several regions not indicated by Kohara that almost completely lack *Eco*RV information. This clustering of missing *Eco*RV sites suggests that individual clone DNA preparations (perhaps residual amounts of DNA purification reagents), rather than context effects, are responsible for these clustered missing sites. Whatever the reason for this inefficient cleavage, Isono has replaced *Eco*RV with *Xho*I for the subsequent restriction mapping of yeast chromosomes III and VI (Yoshikawa and Isono 1990, 1991).

As noted earlier (Churchill et al. 1990; Karlin and Macken 1991b), the efficiency of restriction site detection is lower with the more frequently cutting enzymes, probably due to an increased likelihood that two sites will occur close together. A theoretical relationship between observed efficiency of restriction site detection and site abundance can be derived from equation 12 of Churchill et al. (1990). Let E be the efficiency (as in Table 1), set D = (number of sites)/(DNA length), and let t be minimum spacing below which sites for the same enzyme will be mapped to the same site. Then $1 + tD = 1/E$.

Using observed values for E and D, we found the value of t giving a curve (Fig. 7) that best fits the empirical data, excluding the *Eco*RV and *Bgl*I sites (*Bgl*I has an 11-bp recognition site and is not a palindromic hexamer). This model predicts that sites closer than 763 bp were scored as a single site. Considering that many of the partial restriction fragments whose lengths were estimated on gels (Kohara et al. 1987) were in the poorly resolved 20–30-kb range, this figure is not surprising. We are currently evaluating the reliability of our estimates of site detection efficiency as part of a comprehensive analysis of all the actual MapSearch alignments (to be presented elsewhere).

The distribution of *Eco*RI and *Hin*dIII sites

We determined the percentage of gel-mapped restriction sites (Ecoli2.map) that are contained in sequenced regions (EcoSeq5). Again taking *Bam*HI as an example, 15 (i.e., 488 – 473) of the 194 EcoSeq5 (aligned subset) sites are missing from Ecoli2.map, so EcoSeq5 contains 179 (i.e., 194 – 15) of the 473 Ecoli2.map sites. Thus, 37.8% of the *Bam*HI sites in Ecoli2.map are contained within the 38% of the chromosome that is compiled in EcoSeq5.

It is unclear to us why two enzymes, *Eco*RI and *Hin*dIII, have disproportionately high percentages (46.9% and 47.5%, respectively) of their Ecoli2.map sites in sequenced regions (Table 1). These are the two AT-rich recognition sequences, but the GC-rich sites are partitioned between sequenced and unsequenced fractions of the genome without any large bias in distribution, so it seems unlikely that GC content is the

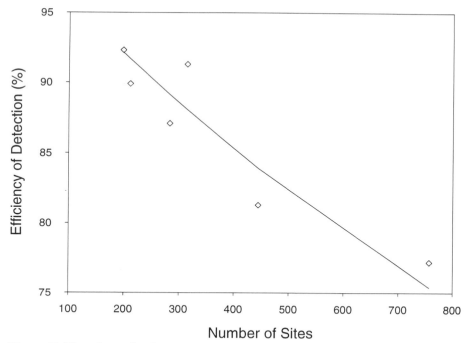

Figure 7 The relationship between restriction site density and the efficiency of detection by the method of Kohara et al. (1987). EcoRV data were omitted because of their incompleteness, and BgII, the only site that is not a palindromic hexamer, was omitted because it gave anomalous results. The best-fit line is derived using an equation presented by Churchill et al. (1990) (see text).

deciding factor. In addition, the sample size (38%) is so large that purely stochastic explanations are unconvincing. Finally, correcting in various ways for the observed levels of missed sites (9% and 13%, respectively) does not explain the 24% discrepancy between the 47% and 38% figures.

Rare-cutting restriction enzyme sites in EcoSeq5

Table 2 lists the numbers of sites in EcoSeq5 for various rare-cutting enzymes. Genomic restriction maps have been proposed for AvrII (Condemine and Smith 1990; Daniels 1990a), NotI (Smith et al. 1987; Heath et al. 1992), XbaI (J.D. Heath et al., in prep.), and SfiI (Condemine and Smith 1990; Perkins et al. 1992). Except for AvrII, these sites have been mapped by hybridizing excised PFG restriction fragments to the Kohara miniset clones. A similar approach using an alternative collection of E. coli MG1655 λ clones was used for AvrII (Daniels 1990a). Further confirmation can be obtained by isolating Kohara miniset clone DNA and demonstrating the presence of a rare restriction site, but this has not been reported.

Table 2 Frequency of selected rare enzyme recognition sites

Enzyme	Recog. Site	Number in EcoSeq5[a]	Total in genome[b]
*Avr*II	C^CTAGG	14	17
*Not*I	GC^GGCCGC	13	22
*Sfi*I	GGCCNNNN^NGGCC	21	31
*Xba*I	T^CTAGA	17	35
*Asc*I	GG^CGCGCC	59	
*Apa*I	GGGCC^C	38	
*Fse*I	GGCCGG^CC	4	
*Nhe*I	G^CTAGC	49	
*Pac*I	TTAAT^TAA	48	
*Pme*I	GTTT^AAAC	31	
*Spe*I	A^CTAGT	40	
*Srf*I	GCCC^GGGC	17	
*Sse*8387I	CCTGCA^GG	31	
*Swa*I	ATTT^AAAT	36	

[a]Two of the *Avr*II sites in EcoSeq5 are suspect (see text).
[b]The total number of sites (not including bacteriophage λ sites) are presented for four enzymes and are from the following sources: *Avr*II (see text); *Not*I (Heath et al. 1992); *Sfi*I (Perkins et al. 1992); *Xba*I (J.D. Heath et al., in prep.).

Studies by Perkins et al. (1992) and Heath et al. (1992 and in prep.) utilize the nonredundant EcoSeq database and the corresponding integrated genomic map to calculate genomic positions for the rare restriction sites. One advantage to this approach is that rare-enzyme site positions can be easily integrated into a single map with the eight Kohara enzymes (Perkins et al. 1992; J.D. Heath et al., in prep.). Another advantage is that all the miniset clones have also been positioned on the integrated genomic map, allowing the sites identified by hybridizing PFG fragments to the Kohara miniset to be positioned on the integrated map. As noted previously (Smith and Condemine 1990; Heath et al. 1992 and in prep.; Perkins et al. 1992), the composite rare restriction site map has numerous practical applications, including the rapid characterization of naturally occurring and mutant chromosomal rearrangements such as inversions, duplications, deletions, and translocations.

Positions of sequenced rare restriction sites and the sequenced portions of the *E. coli* chromosome are displayed in Figure 6. Only four *Fse*I sites have been sequenced in *E. coli* so far, making it the rarest of the 6-base or 8-base rare cutters. With the abundant DNA sequence information available for *E. coli*, application of the recently developed "Achilles' heel" technique (Grimes et al. 1990; Koob and Szybalski 1990) would allow essentially any small set of sequenced restriction sites to be the only sites cleaved. In this technique, the DNA sequences around a site or set of sites are synthesized and added to a genomic DNA sample along with the RecA protein. Subsequent treatment with a methylase (followed by deproteinization and restriction enzyme digestion) will protect all the restriction sites from cleavage by the companion restriction enzyme except those sites covered by the RecA-DNA complex.

Care must be taken when attempting to correlate sequenced restriction sites with observed cleavage sites. For example: The *NotI* site in the *bioA* gene sequence (ECOBIO; J04423) cannot be correlated to an actual site and could be a sequencing error (Medigue et al. 1991; Heath et al. 1992); several *SfiI* sites overlap with *Dcm* methylation sites, which in some cases can interfere with *SfiI* digestion (Nelson and McClelland 1991; Perkins et al. 1992); the *AvrII* site reported to be in the IS5 DNA sequence (Condemine and Smith 1990) is actually not in the IS5 sequence, but is a hybrid site created when IS5 transposed into cloned *Haemophilus* DNA (Schoner and Kahn 1981; Daniels 1990a); the *XbaI* site in the *sbcB* gene sequence overlaps and is blocked by a *Dam* methylation site (Nelson and McClelland 1991; J.D. Heath et al., in prep.); and the *XbaI* site in IS30 is not present at all chromosomal IS30 locations since they have not all been sequenced and some are partial copies (J.D. Heath et al., in prep.). Strain differences have been observed among the rare restriction sites (Daniels 1990a; Heath et al. 1992 and in prep.; Perkins et al. 1992) and may account for some of the DNA sequence discrepancies.

AvrII map of E. coli

Information extracted from EcoSeq5 allows us to predict sizes for most of the chromosomal *AvrII* restriction fragments, permitting comparison with earlier *AvrII* maps of several different strains of *E. coli* (Daniels 1990a; Condemine and Smith 1990). Direct comparison to strain MG1655 is possible, since the Kohara map for strain W3110 has been modified to return the IN(*rrnD-E)1* inversion to the MG1655 (wild type) configuration (Rudd et al. 1990). In Figure 8, the chromosomal *AvrII* sites determined by Daniels are compared with the EcoSeq5 predictions, and in Table 3, the correlation of these sites is summarized. We note that the cumulative length of the *E. coli* chromosome estimated by Daniels (4835 kb) is close to our estimate based on Ecoli5.map (4673.6 kb).

We predict 17 chromosomal *AvrII* sites, 4 more than reported by Daniels (1990a). Daniels did not account for the sites in the glutamate tRNA genes: *gltW, gltU, gltT,* and *gltV* (Harvey et al. 1988; Condemine and Smith 1990). She noted the *gltT* site as a second *AvrII* site in the

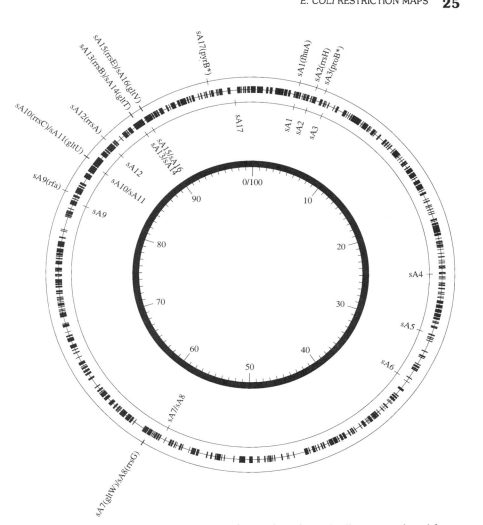

Figure 8 Comparison of the *Avr*II map of Daniels with an *Avr*II map produced from Ecoli5.map and EcoSeq5. (*Outer circle*) Sites in EcoSeq5 pinned to Ecoli5.map (see Table 3 for explanation of the sites with an asterisk). (*Next circle inward*) The sequenced portions of Ecoli5.map. (*Next circle inward*) The positions of the *Avr*II sites taken from the map of Daniels (1990a). All of Daniels' fragment sizes were multiplied by 0.96562 to bring the two maps into alignment since the sum of the fragments was slightly larger than the size of the *E. coli* genome as estimated by Ecoli5.map. (*Inner circle*) A genomic scale based on centisomes, which approximate genetic map minutes.

rrnB operon, but it was not counted among her 13 sites. The presence of these additional *Avr*II sites is obscured during pulsed-field gel electrophoresis (PFGE) because they are only 1.5 kb away from sites in the 16S rRNA genes: *rrsG, rrsC, rrsB,* and *rrsE.* Only *rrsG* and *rrsB* DNA

Table 3 Chromosomal AvrII restriction sites

Site[a]	Length(D)[b]	Length(E)[c]	Gene(D)	Gene(E)	DNA Sequence
sA1	55-b	63.1	fhuA	fhuA	ECOFHUACD
sA2	70-c	35.7	phoE	rrsH	ECORGNB[d]
sA3	900-j1(2480)	(2472)	n.d.	proB*	ECOPHOE*
sA4	255-h2	n.d.	n.d.	n.d.	n.d.
sA5	215-g	n.d.	n.d.	n.d.	n.d.
sA6	1110-j2	n.d.	n.d.	n.d.	n.d.
sA7	n.d.	1.5	rrnG	gltW	M20397
sA8	1150-j3	1080	rrnG	rrsG	ECOPROT
sA9	170-f	148.5	n.d.	rfa	ECORFA2
sA10	n.d.	1.4	oriC	rrsC	ECORGNB[d]
sA11	105-d	93.7	oriC	gltU	ECO16S23S
sA12	145-e	131.0	rrnC	rrsA	ECORGNB[d]
sA13	n.d.	1.5	rrnB	rrsB	ECORGNB
sA14	35-a	40.9	rrnB	gltT	ECORGNB
sA15	n.d.	1.4	rrnE	rrsE	ECORGNB[d]
sA16	375-i	262.1	rrnE	gltV	gltVeco[e]
sA17	255-h1	340.5	pyrB	pyrB*	ECOPYRBIA*

The proB and pyrB sites are marked with an asterisk because the sequence data are ambiguous and they may not correspond to sites sA3 and sA17, respectively. n.d indicates not determined.

[a] Sites are named using the convention set by the Weinstock group (Heath et al. 1992 and in prep.; Perkins et al. 1992).

[b] D denotes a fragment length (starting with the indicated site) as determined by Daniels (1990a). Parenthetical lengths are cumulative for the sA3–sA7 fragment.

[c] E denotes a fragment size predicted using the EcoSeq5 data set.

[d] The ECORGNB rrsB sequence is used as an analog for the rrsH, rrsC, rrsA, and rrsE sequences.

[e] The gltVeco DNA sequence was taken from Harvey et al. (1988).

sequences have been published, and both contain *Avr*II sites. DNA sequencing has recently revealed that the *rrsA* gene also has an *Avr*II site (D. Daniels, pers. comm.).

The sites predicted using EcoSeq5 compare well to the Daniels sites with a few exceptions. Sites sA4–6 do not have corresponding sites in EcoSeq5, but the predicted and observed sizes of the sA3–sA7 fragment are close (2472 to 2480 kb, Table 3). The putative sequenced sA17 site is not very close to the Daniels sA17 *Avr*II site and could be a DNA sequencing error. The identification of the sequenced sA3 site is also quite tentative due to ambiguous DNA sequencing data (see below).

The *Avr*II analysis revealed some possible sequencing errors that complicate the correlation of the Daniels and EcoSeq5 sites. The *Avr*II site in the *ebgA* DNA sequence noted by Daniels (1990a) has since been corrected as a DNA sequencing error (B. Hall, pers. comm.; GenBank ECOEBGRA) and was not found by PFGE (Daniels 1990a). The DNA downstream from the *nusB* and *glgX* genes (ECONUSAA and ECOGLG) are predicted to have *Avr*II sites, but these sites were not confirmed by PFGE; they are likely to be DNA sequencing errors or sites that are not cut efficiently. The *oriC* region of *E. coli* has been sequenced several times, and only one of the versions in GenBank has the *Avr*II site noted by Daniels (1990a; Medigue et al. 1991). We consider this site to be an artifact (Table 3). The DNA at the *pyrB* *Avr*II site (sA17) has been sequenced twice (ECOPYRBI and ECOPYRBIA), but only ECOPYRBIA contains an *Avr*II site. Although this is somewhat near sA17, confirmation of the *Avr*II site in *pyrB* requires either more DNA sequencing or mapping experiments similar to those reported by Heath et al. (1992). Likewise, the sequenced *Avr*II site close to sA3 that is in the *proB* gene DNA sequence suffers from not being present in one of the two sources of sequence data for the *proB* gene (ECOPHOE has the site and ECOPHOEA does not). These two tentatively correlated *Avr*II sites derived from ambiguous sequence data are marked with an asterisk in Figure 8. We note that the *Avr*II site identified in the EMBL entry ECSPROT (X52620) by Medigue et al. (1991) is an artifact that arises because the ECSPROT entry erroneously contains 3′–5′ DNA sequence information; this has been corrected in the EcoSeq database, but not in EMBL or GenBank.

The frequency distribution of palindromic hexamers

Table 4 and Figure 9 give the frequencies with which all 64 palindromic hexamers occur in EcoSeq5, as computed by the Prolog query system discussed above. Restriction enzymes that recognize 52 of these sites have been discovered, and their names and recognition/cleavage sites are presented in Table 4. In Table 4, two measures of observed-to-predicted frequency bias are also reported. The first is a simple measure based on

Table 4 Palindromic hexamers of *E. coli*

Enz.	Site	EcoSeq5	Simple	Markov
*Avr*II	C^CTAGG	17	0.04	1.19
*Xba*I	T^CTAGA	17	0.04	1.85
*Apa*I	GGGCC^C	38	0.09	0.26
*Spe*I	ACTAGT	40	0.09	1.18
*Nhe*I	GCTAGC	49	0.11	1.11
*Nar*I	GG^CGCC	62	0.14	0.08
*Sac*I	GAGCT^C	70	0.16	0.32
*Xho*I	C^TCGAG	74	0.17	0.51
*Pma*CI	CAC^GTG	84	0.19	0.34
n.d.	CTATAG	93	0.22	0.86
n.d.	TAGCTA	119	0.28	1.48
n.d.	TATATA	134	0.31	1.22
*Xma*III	C^GGCCG	156	0.36	0.41
*Nae*I	GCC^GGC	157	0.36	0.22
*Sca*I	AGT^ACT	159	0.37	0.63
*Sna*I	GTATAC	162	0.37	1.06
*Sma*I	CCC^GGG	192	0.44	0.52
n.d.	ACATGT	194	0.45	0.92
*Bam*HI	G^GATCC	198	0.46	0.59
*Sna*BI	TAC^GTA	202	0.47	0.94
*Kpn*I	GGTAC^C	211	0.49	0.58
*Sph*I	GCATG^C	212	0.49	0.48
*Afl*II	C^TTAAG	216	0.50	1.13
n.d.	ATATAT	219	0.51	0.98
*Sal*I	G^TCGAC	223	0.52	0.52
n.d.	TGTACA	232	0.54	0.99
*Apa*LI	G^TGCAC	233	0.54	0.78
*Nde*I	CATATG	252	0.58	0.93
*Nco*I	C^CATGG	259	0.60	0.66
*Spl*I	C^GTACG	259	0.60	0.62
*Stu*I	AGG^CCT	263	0.61	0.73
*Sac*II	CCGC^GG	267	0.62	0.49

Table 4 (continued)

Enz.	Site	EcoSeq5	Simple	Markov
*Bal*I	TGG^CCA	278	0.64	0.61
*Hind*III	A^AGCTT	283	0.65	0.77
*Ava*III	ATGCAT	289	0.67	0.78
*Aat*II	GACGT^C	300	0.69	0.82
*Bgl*II	A^GATCT	309	0.71	0.91
*Eco*RI	G^AATTC	312	0.72	0.85
n.d.	TTATAA	316	0.73	0.98
*Eco*47III	AGC^GCT	324	0.75	0.47
*Mfe*I	C^AATTG	329	0.76	1.03
n.d.	TAATTA	329	0.76	1.06
n.d.	TTGCAA	336	0.78	0.77
*Asu*II	TT^CGAA	375	0.87	0.98
*Bsp*MII	T^CCGGA	410	0.95	0.60
*Pst*I	CTGCA^G	444	1.03	0.55
*Bsp*HI	T^CATGA	446	1.03	0.94
*Pvu*I	CGAT^CG	525	1.21	0.83
*Mlu*I	A^CGCGT	527	1.22	0.94
*Cla*I	AT^CGAT	544	1.26	0.75
*Vsp*I	AT^TAAT	560	1.30	0.95
n.d.	AAATTT	566	1.31	0.74
*Aha*III	TTT^AAA	574	1.33	1.02
*Nru*I	TCG^CGA	576	1.33	0.99
*Ssp*I	AATATT	624	1.44	0.91
*Bcl*I	T^GATCA	628	1.45	0.98
*Hpa*I	GTT^AAC	633	1.46	1.05
n.d.	AACGTT	678	1.57	0.83
*Pvu*II	CAG^CTG	757	1.75	0.75
*Age*I	ACCGGT	761	1.76	1.06
*Mst*I	TGC^GCA	771	1.78	0.82
*Eco*RV	GAT^ATC	797	1.84	1.15
n.d.	CGCGCG	881	2.04	0.85
*Bse*PI	GCGCGC	1017	2.35	0.82

See text for an explanation of the simple and second order Markov chain observed/expected bias calculations. n.d. indicates not determined.

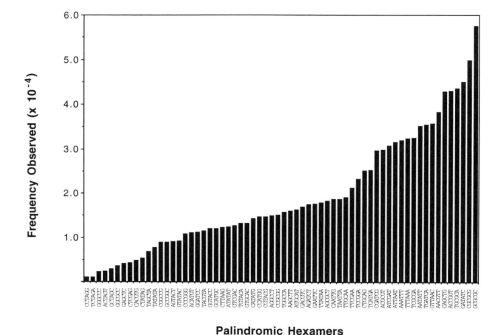

Palindromic Hexamers

Figure 9 Frequency distribution of all 64 palindromic hexamers in EcoSeq5.

the random expected hexamer frequency of 1/4096, or 432.3 sites in EcoSeq5. The second measure, calculated using second order Markov chains (Phillips et al. 1987a,b), adjusts each hexamer frequency expectation using the frequencies observed for the two pentamer subsequences. Since each oligomer's observed frequency will reflect nonrandom bias in any of the lower oligomer classes, the difference between the simple and Markov predictions reflects nonrandomness in subsequence oligomers. The 14 hexamers with Markov bias values less than 0.6 are all GC-rich. We note that the mononucleotide frequencies show no significant strand bias: The G+C content is 51.52%, with 24.26% A, 25.70% C, 25.82% G, and 24.22% T.

The four palindromic hexamers containing CTAG are among the five rarest. The Markov bias values confirm subsequence (i.e., CTAG) underrepresentation. This underrepresentation of the tetramer CTAG has been previously noted and discussed (McClelland et al. 1987; Phillips et al. 1987a,b; Medigue et al. 1991; Burge et al. 1992; McClelland and Bhagwat 1992; Karlin et al. 1992). One explanation for a low CTAG frequency in *E. coli* could be the action of the *vsr* gene product to repair away the sequence CTAG (and others) when it is involved in a DNA mismatch (Hennecke et al. 1991), which could occur as a DNA replication error as well. McClelland and Bhagwat (1992), using data obtained from EcoSeq5 using a Prolog query, report that the targets of the Vsr enzyme, including CTAG, are underrepresented, whereas the tetramers resulting from Vsr repair are overrepresented. Structural kinking of DNA at CTAG sequences has also been proposed as a contributing factor (Medigue et al. 1991; Burge et al. 1992), which may help explain why all DNA is low in CTAG. Table 4 lists several other enzymes that could be useful as rare cutters in *E. coli*, particularly *Apa*I (GGGCCC).

The hexamers CGCGCG and GCGCGC are the most frequent. The Markov bias values indicate that the subsequences are also over-represented. As pointed out by Karlin et al. (1992), the iterative nature of these two sequences leads to their being overcounted as independent occurrences when they appear in, e.g., the octamer CGCGCGCG (85 of which appear in Ecoli5.seq) or the 11-mer CGCGCGCGCGCG (one copy). However, even taking their iterative nature into account, these sequences are still overrepresented (Karlin et al. 1992). On the other hand, the other repeating-dinucleotide hexamers, ATATAT and TATATA, are underrepresented and have very few of the corresponding higher order octamers (9 each).

Comparison of the genetic and genomic restriction maps of *E. coli*

Alignment of DNA sequences to a genomic restriction map permits comparison of gene positions on a genetic map with positions on a genomic restriction map (Kohara et al. 1987; Bachmann 1990; Kohara 1990; Medigue et al. 1990a,b, 1991; Rudd et al. 1990, 1991; Watanabe and Kunisawa, 1990; Rudd 1992). A previous comparison of *E. coli* maps noted a smooth departure from strict colinearity (cocircularity?) in a large region from 40 to 80 minutes on the genetic map (Rudd et al. 1990). The residual, a measure of the difference between genetic and restriction maps, decreased when the comparison was made to successive genetic map updates (Bachmann et al. 1976; Bachmann and Low 1980; Bachmann 1983). Since that original report, both the genetic map (Bachmann 1990) and the genomic restriction map (Rudd et al. 1991; Rudd 1992) have been revised. A total of 51.3 kb has been removed from an earlier digital version of the *E. coli* genomic restriction map at a position near 70 minutes, as described above.

We realigned a sample of 1990 genetic map positions to Ecoli5.map (Fig. 10) and saw little effect on the magnitude of the residual. A possible source of this persistent discrepancy could be located in the region around minute 65 of the 1990 *E. coli* genetic map. The 1.4-minute interval between *metC* (65.0 min) and *tolC* (66.4 min) can be converted (46.7 kb/min) to an estimated physical separation distance of 65.4 kb. However, we predict that *metC* and *tolC* are separated

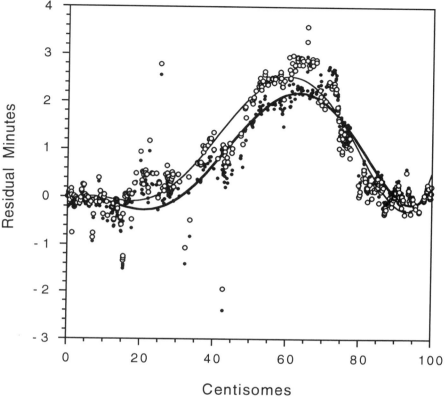

Figure 10 Comparison of the *E. coli* genetic and genomic restriction maps. The effect of removing minute 65 from the genetic map is shown (see text). Gene positions: before removal (*open circle*); after removal (*filled circle*).

by only 25.8 kb using the EcoGene5 map coordinates (Rudd 1992; Rudd and Schneider 1992). There are no accurately mapped genes between 65.0 and 66.0 minutes on the 1990 genetic map. The discrepancy between the genetic and physical map estimates of the *metC-tolC* interval, 39.6 kb, is close to the value of a minute, so we explored the effect of removing one minute (65–66 min) from the 1990 genetic map (Fig. 10). The effect on the magnitude of the residual "hump" is slight, but generally a better alignment is achieved. The *tolC* region of the *E. coli* chromosome may need to be reinvestigated using genetic mapping techniques (Hfr mating and P1 cotransduction) and physical mapping (e.g., clone hybridization) to explore possible explanations for the mapping discrepancies. The persistent "hump" deviation may have a variety of sources, including an endpoint effect due to the use of a linear plot to depict a circular map.

Using MapSearch to locate repeated DNA in *E. coli*

We have previously described the detection of moderate-size copies of repeat DNA, such as four copies of a 14-kb fragment containing the *tdc* operon (Rudd et al. 1990, 1991), the seven copies of the 23S and 16S rRNA genes (Rudd et al. 1990), a possible second copy of the *fec* operon (Rudd et al. 1991), and the *rhs* duplicate genes (Rudd et al. 1991). When mapping the IS elements of *E. coli* using MapSearch (Rudd et al. 1990, 1991; Rudd 1992) and Kohara clone hybridization data (Umeda and Ohtsubo 1989, 1990a,b; Birkenbihl and Vielmetter 1989a, 1991), we were surprised to find that the small (~1.3 kb) repeat DNA of the IS elements could also be detected as repeated MapSearch hits. In Table 5 are listed the positions of the IS186, IS2, and IS3 sequences that were located using a single IS DNA sequence as a Probe. Thus, even two-site Probes, effective in locating some nonrepeated DNA sequences (Rudd et al. 1991), can be used to detect small segments of repeated DNA.

CONCLUSION

We have described a computer system that uses a genomic restriction map as the basis for organizing information about *E. coli*. This approach was inspired by the longstanding use of the *E. coli* genetic map to organize and access information about *E. coli*, particularly as a source for references to the primary literature. We offer the integrated genomic map and the EcoSeq nonredundant DNA sequence collection to the *E. coli* research community for use as a tool to expedite completion of the *E. coli* genome sequence.

Our original goals were to organize all known *E. coli* DNA sequences into a nonredundant data set and to integrate these sequences with the genomic restriction map. The decision was made to collect these data into plain text files, which simplifies communicating the data to others and avoids making any assumptions about how the data should best be organized. Subsequently, we extended coverage to include gene-coding regions and provided users with a graphical interface to the data sets (GeneScape). Our data sets supply the grist for several research projects, by ourselves and other workers, to investigate appropriate methods of data organization and presentation. Some of these projects are discussed above.

We are continuing to expand our coverage of *E. coli* data to include enzyme names, synonyms, protein homologs, EC numbers, pathways, and many references. We envision the genomic restriction map and the genes and DNA sequences aligned to it as the backbone for a larger database ultimately aimed at modeling the structure and function of the *E. coli* cell. Moreover, we have used *E. coli* as a model system to develop new methods that should be useful for other genome bioin-

Table 5 MapSearch alignments of IS186, IS3, and IS2

IS[a]	Orient.	Min.[b]	Rank	p Value	Bp	Sites
IS186A	+	0.300	1	0.326	1336	2
IS186B	+	14.100	1	0.326	1336	2
IS186C	+	51.800	3	1.000	1336	2
IS2A	−	8.400	11	0.975	1331	3
IS2B	+	23.300	2	0.385	1331	3
IS2C	−	27.400	4	0.888	1331	3
IS2D	+	31.000	3	0.432	1331	3
IS2F	+	43.400	13	1.000	1331	3
IS2G	+	47.500	9	0.959	1331	3
IS2H	+	61.500	12	0.951	1331	3
IS2I	−	66.500	15	1.000	1331	3
IS2K	+	97.050	19	1.000	1331	3
IS3A	+	7.000	4	0.783	1258	4
IS3B	−	8.600	25	1.000	1258	4
IS3C	+	12.400	10	0.994	1258	4
IS3D	−	23.100	3	0.772	1258	4
IS3E	+	44.300	1	0.227	1258	4

[a]The IS186 Probe was derived from the GenBank entry INS186ECB (X03123); the IS2 Probe was derived from bp 66–1396 of the GenBank entry ECOINS2K (V00279); the IS3 Probe was derived from Genbank entry INS3 (X02311). The IS2 and IS3 positions were identified by comparing the IS containing miniset clones identified by Umeda and Ohtsubo (1989) with those predicted from the MapSearch alignments. The IS2 (A–K) and IS3 (A–E) designations are taken from Umeda and Ohtsubo (1989). The IS186 assignments were confirmed using the data of Birkenbihl and Vielmetter (1991).
[b]The map positions in minutes are taken from Umeda and Ohtsubo (1989) and Birkenbihl and Vielmetter (1991).

formatics projects. We expect that many other organisms, both prokaryotic and eukaryotic, will soon have the wealth of genome-related data currently available for *E. coli*.

Acknowledgments

We thank Ross Overbeek for the Prolog implementation of the *E. coli* genomic data sets and John Wilbur for assistance in the theoretical analysis of closely spaced restriction sites. We also thank everyone who has assisted in various aspects of this project, including D. Benson, S. Karlin, D. Kingsbury, J. Ostell, M. Riley, S. Satterfield, M. Shapiro, K. Sirotkin, C. Tolstoshev, and C. Werner. We also thank G. Weinstock for providing unpublished data and for helpful comments on the *Avr*II computer analysis. Last, we extend our appreciation to all those members of the *E. coli* research community who have shared their data and ideas with us. W. M. was supported in part by grant ROI LM-05110 from the

National Library of Medicine. The work of G.B. was supported in part by grant DIR-8902282 from the National Science Foundation.

References

Altschul, S.F., W. Gish, W. Miller, E.W. Myers, and D.J. Lipman. 1990. Basic local alignment search tool. *J. Mol. Biol.* **215**: 403.

Bachmann, B.J. 1983. Linkage map of *Escherichia coli* K-12, edition 7. *Microbiol. Rev.* **47**: 180.

———. 1990. Linkage map of *Escherichia coli* K-12, edition 8. *Microbiol. Rev.* **54**: 130.

Bachmann, B.J. and K.B. Low. 1980. Linkage map of *Escherichia coli* K-12, edition 6. *Microbiol. Rev.* **44**: 1.

Bachmann, B.J., K.B. Low, and A.L. Taylor. 1976. Recalibrated linkage map of *Escherichia coli* K-12. *Bacteriol. Rev.* **40**: 116.

Berlyn, M. and S. Letovsky. 1992. COTRANS: A program for co-transduction analysis. *Genetics* (in press).

Birkenbihl, R.P. and W. Vielmetter. 1989a. Complete maps of IS1, IS2, IS3, IS4, IS5, IS30 and IS150 locations in *Escherichia coli* K12. *Mol. Gen. Genet.* **220**: 147.

———. 1989b. Cosmid-derived map of *E. coli* strain BHB2600 in comparison to the map of strain W3110. *Nucleic Acids Res.* **17**: 5057.

———. 1991. Completion of the IS map in *E. coli*: IS186 positions on the *E. coli* K12 chromosome. *Mol. Gen. Genet.* **226**: 318.

Bouffard, G., J. Ostell, and K.E. Rudd. 1992. GeneScape: A relational database of *Escherichia coli* genomic map data for Macintosh computers. *Comput. Appl. Biosci.* (in press).

Burge, C., A.M. Campbell, and S. Karlin. 1992. Over- and under-representation of short oligonucleotides in DNA sequences. *Proc. Natl. Acad. Sci.* **89**: 1358.

Churchill, G.A., D.L. Daniels, and M.S. Waterman. 1990. The distribution of restriction enzyme sites in *Escherichia coli*. *Nucleic Acids Res.* **18**: 589.

Condemine, G. and C.L. Smith. 1990. Genetic mapping using large-DNA technology: Alignment of *Sfi*I and *Avr*II sites with the *Not*I genomic restriction map of *Escherichia coli* K-12. In *The bacterial chromosome* (ed. K. Drlica and M. Riley), p. 53. American Society for Microbiology, Washington, D.C.

Conway, T., K.C. Yi, S.E. Egan, R.E. Wolf, Jr., and D.L. Rowley. 1991. Locations of the *zwf*, *edd*, and *eda* genes on the *Escherichia coli* physical map. *J. Bacteriol.* **173**: 5247.

Danchin, A., C. Medigue, O. Gascuel, H. Soldano, and A. Henault. 1991. From data banks to databases. *Res. Microbiol.* **142**: 913.

Daniels, D.L. 1990a. The complete *Avr*II restriction map of the *Escherichia coli* genome and comparisons of several laboratory strains. *Nucleic Acids Res.* **18**: 2649.

———. 1990b. Constructing encyclopedias of genomes. In *The bacterial chromosome* (ed. K. Drlica and M. Riley), p. 43. American Society for Microbiology, Washington, D.C.

Daniels, D.L. and F.R. Blattner. 1987. Mapping using gene encyclopaedias. *Nature* **325**: 831.

Grimes, E., M. Koob, and W. Szybalski. 1990. Achilles' heel cleavage: creation of rare restriction sites in lambda phage genomes and evaluation of additional operators, repressors and restriction/modification systems. *Gene* **90**: 1.

Gumbel, E.J. 1962. Statistical theory of extreme values (main results). In *Contributions to order statistics* (ed. A.E. Sarhan and B.G. Greenberg), p. 56. John Wiley, New York.

Harvey, S., C.W. Hill, C. Squires, and C.L. Squires. 1988. Loss of spacer loop sequence from the *rrnB* operon in the *Escherichia coli* K-12 subline that contains the *relA1* mutation. *J. Bacteriol.* **170**: 1235.

Heath, J.D., J.D. Perkins, B. Sharma, and G.M. Weinstock. 1992. *Not*I cleavage map of *Escherichia coli* K-12 strain MG1655. *J. Bacteriol.* **174**: 558.

Hennecke, F., H. Kolmar, K. Brundl, and H.J. Fritz. 1991. The *vsr* gene product of *E. coli* K-12 is a strand- and sequence-specific DNA mismatch repair endonuclease. *Nature* **353**: 776.

Huang, X. and M.S. Waterman. 1992. Dynamic programming algorithms for restriction map comparison. *Comput. Appl. Biosci.* (in press).

Karlin, S. and C. Macken. 1991a. Some statistical problems in the assessment of inhomogeneities of DNA sequence data. *J. Am. Statist. Assoc.* **86**: 27.

―――. 1991b. Assessment of inhomogeneities in an *E. coli* physical map. *Nucleic Acids Res.* **19**: 4241.

Karlin, S., C. Burge, and A.M. Campbell. 1992. Statistical analyses of counts and distributions of restriction sites in DNA sequences. *Nucleic Acids Res.* **20**: 1363.

Knott, V., D.J. Blake, and G.G. Brownlee. 1989. Completion of the detailed restriction map of the *E. coli* genome by the isolation of overlapping cosmid clones. *Nucleic Acids Res.* **17**: 5901.

Knott, V., D.J. Rees, Z. Cheng, and G.G. Brownlee. 1988. Randomly picked cosmid clones overlap the *pyrB* and *oriC* gap in the physical map of the *E. coli* chromosome. *Nucleic Acids Res.* **16**: 2601.

Kohara, Y. 1990. Correlation between the physical and genetic maps of the *Escherichia coli* K-12 chromosome. In *The bacterial chromosome* (ed. K. Drlica and M. Riley), p. 29. American Society for Microbiology, Washington, D.C.

Kohara, Y., K. Akiyama, and K. Isono. 1987. The physical map of the whole *E. coli* chromosome: Application of a new strategy for rapid analysis and sorting of a large genomic library. *Cell* **50**: 495.

Komine, Y., T. Adachi, H. Inokuchi, and H. Ozeki. 1990. Genomic organization and physical mapping of the transfer RNA genes in *Escherichia coli* K12. *J. Mol. Biol.* **212**: 579.

Koob, M. and W. Szybalski. 1990. Cleaving yeast and *Escherichia coli* genomes at a single site. *Science* **250**: 271.

Kroger, M., R. Wahl, and P. Rice. 1991. Compilation of DNA sequences of *Escherichia coli* (update 1991). *Nucleic Acids Res.* (supp.) **19**: 2023.

Kunisawa, T. and M. Nakamura. 1991. Identification of regulatory building blocks in *Escherichia coli* genome. *Protein Seq. Data. Anal.* **4**: 43.

Kunisawa, T., M. Nakamura, H. Watanabe, J. Otsuka, A. Tsugita, L.S. Yeh, and D.G. George. 1990. *Escherichia coli* K12 genomic database. *Protein Seq. Data. Anal.* **3**: 157.

Lipinska, B., O. Fayet, L. Baird, and C. Georgopoulos. 1989. Identification, characterization, and mapping of the *Escherichia coli htrA* gene, whose product is essential for bacterial growth only at elevated temperatures. *J. Bacteriol.* 171: 1574.

McClelland, M. and A.S. Bhagwat. 1992. Biased DNA repair. *Science* **355**: 595.

McClelland, M., R. Jones, Y. Patel, and M. Nelson. 1987. Restriction endonucleases for pulsed field mapping of bacterial genomes. *Nucleic Acids Res.* **15**: 5985.

Medigue, C., A. Henaut, and A. Danchin. 1990a. *Escherichia coli* molecular genetic map (1000 kbp): Update I. *Mol. Microbiol.* **4**: 1443.

Medigue, C., J.P. Bouche, A. Henaut, and A. Danchin. 1990b. Mapping of sequenced genes (700 kbp) in the restriction map of the *Escherichia coli* chromosome. *Mol. Microbiol.* **4**: 169.

Medigue, C., A. Viari, A. Henault, and A. Danchin. 1991. *Escherichia coli* molecular genetic map (1500 kbp): Update II. *Mol. Microbiol.* **5**: 2629.

Miller, W., J. Barr, and K.E. Rudd. 1991. Improved algorithms for searching restriction maps. *Comput. Appl. Biosci.* **7**: 447.

Miller, W., J. Ostell, and K.E. Rudd. 1990. An algorithm for searching restriction maps. *Comput. Appl. Biosci.* **6**: 247.

Nelson, M. and M. McClelland. 1991. Site-specific methylation: Effect on DNA modification methyltransferases and restriction endonucleases. *Nucleic Acids Res.* (suppl.) **19**: 2045.

Oh, B.K., A.K. Chauhan, K. Isono, and D. Apirion. 1990. Location of a gene (*ssrA*) for a small, stable RNA (10Sa RNA) in the *Escherichia coli* chromosome. *J. Bacteriol.* **172**: 4708.

Pearson, W.R. and D.J. Lipman. 1988. Improved tools for biological sequence comparison. *Proc. Natl. Acad. Sci.* **85**: 2444.

Perkins, J.D., J.D. Heath, B.R. Sharma, and G.M. Weinstock. 1992. *Sfi*I genomic cleavage map of *Escherichia coli* K-12 strain MG1655. *Nucleic Acids Res.* (in press).

Phillips, G.J., J. Arnold, and R. Ivarie. 1987a. The effect of codon usage on the oligonucleotide composition of the *E. coli* genome and identification of over- and underrepresented sequences by Markov chain analysis. *Nucleic Acids Res.* **15**: 2627.

———. 1987b. Mono- through hexanucleotide composition of the *Escherichia coli* genome: A Markov chain analysis. *Nucleic Acids Res.* **15**: 2611.

Roberts, R.J. and D. Macelis. 1991. Restriction enzymes and their isoschizomers. *Nucleic Acids Res.* (suppl.) **19**: 2077.

Rudd, K.E. 1992. Alignment of *E. coli* DNA sequences to a revised, integrated genomic restriction map. In *A short course in bacterial genetics: A laboratory manual and handbook for* Escherichia coli *and related bacteria* (by J.H. Miller), p. 2.3. Cold Spring Harbor Laboratory Press, Cold Spring Harbor, New York.

Rudd, K.E. and T.D. Schneider. 1992. Compilation of *E. coli* ribosome binding sites. In *A short course in bacterial genetics: A laboratory manual and handbook for* Escherichia coli *and related bacteria* (by J.H. Miller), p. 17.19. Cold Spring Harbor Laboratory Press, Cold Spring Harbor, New York.

Rudd, K.E., W. Miller, J. Ostell, and D.A. Benson. 1990. Alignment of *Escherichia coli* K12 DNA sequences to a genomic restriction map. *Nucleic Acids Res.* **18**: 313.

Rudd, K.E., W. Miller, C. Werner, J. Ostell, C. Tolstoshev, and S.G. Satterfield.

1991. Mapping sequenced *E. coli* genes by computer: Software, strategies and examples. *Nucleic Acids Res.* **19**: 637.

Schoner, B. and M. Kahn. 1981. The nucleotide sequence of IS5 from *Escherichia coli. Gene* **14**: 165.

Shin, D.G., C. Lee, J. Zhang, K.E. Rudd, and C.M. Berg. 1992. Redesigning, implementing and integrating *Escherichia coli* genome software tools with an object-oriented database system. *Comput. Appl. Biosci.* (in press).

Smith, C.L. and G. Condemine. 1990. New approaches for physical mapping of small genomes. *J. Bacteriol.* **172**: 1167.

Smith, C.L., J. Econome, A. Schutt, S. Klco, and C.R. Cantor. 1987. A physical map of the *E. coli* K12 genome. *Science* **236**: 1448.

Taylor, J.D. and S.E. Halford. 1992. The activity of the *Eco*RV restriction endonuclease is influenced by flanking DNA sequences both inside and outside the DNA-protein complex. *Biochemistry* **31**: 90.

Umeda, M. and E. Ohtsubo. 1989. Mapping of insertion elements IS1, IS2 and IS3 on the *Escherichia coli* K-12 chromosome. Role of the insertion elements in formation of Hfrs and F′ factors and in rearrangement of bacterial chromosomes. *J. Mol. Biol.* **208**: 601.

———. 1990a. Mapping of insertion element IS5 in the *Escherichia coli* K-12 chromosome. Chromosomal rearrangements mediated by IS5. *J. Mol. Biol.* **213**: 229.

———. 1990b. Mapping of insertion element IS30 in the *Escherichia coli* K12 chromosome. *Mol. Gen. Genet.* **222**: 317.

VanBogelen, R.A. and F.C. Neidhardt. 1991. The gene-protein database of *E. coli*: Edition 4. *Electrophoresis* **12**: 955.

Watanabe, H. and T. Kunisawa. 1990. Computer-assisted analysis of chromosomal locations and transcriptional directions of *Escherichia coli* genes. *Protein Seq. Data. Anal.* **3**: 149.

Waterman, M.S., T.F. Smith, and H.L. Katcher. 1984. Algorithms for restriction map comparisons. *Nucleic Acids Res.* **12**: 237.

Yoshikawa, A. and K. Isono. 1990. Chromosome III of *Saccharomyces cerevisiae*: An ordered clone bank, a detailed restriction map and analysis of transcripts suggest the presence of 160 genes. *Yeast* **6**: 383.

———. 1991. Construction of an ordered clone bank and systematic analysis of the whole transcripts of chromosome VI of *Saccharomyces cerevisiae. Nucleic Acids Res.* **19**: 1189.

Genome Map of *Drosophila melanogaster* Based on Yeast Artificial Chromosomes

Daniel L. Hartl

Department of Genetics
Washington University School of Medicine
St. Louis, Missouri 63110-1095

A strategy for creating a physical map of the genome of *D. melanogaster* has been successfully implemented. The method is based on DNA fragments of approximately 200 kilobase pairs (kb), cloned in yeast artificial chromosomes (YACs), which have been assigned to positions in the *Drosophila* genome by in situ hybridization with the polytene salivary gland chromosomes. Approximately 90% of the euchromatic genome is included in 1193 YAC clones that have been mapped. The euchromatic genome refers to the banded arms of the polytene chromosomes, which represents about two-thirds of the total genomic DNA, including most of the genes in the organism. The remaining one-third of the genome is heterochromatin, consisting primarily of the Y chromosome and repetitive DNA sequences surrounding the centromeres, which does not become polytene in the salivary gland chromosomes.

Specific topics discussed include:

❏ The reliability of the YAC cloning system in *Drosophila* was evaluated initially by screening libraries with a limited number of single-copy or repetitive DNA sequences. The clones identified with single-copy probes appeared to be genetically stable in the sense that rearrangements do not occur at frequencies high enough to be detected during routine subcloning in yeast. YAC clones containing complex repetitive sequences from heterochromatin were also recovered, including clones containing ribosomal DNA, the telomeric and

Genome Analysis Volume 4: *Strategies for Physical Mapping*
© 1992 Cold Spring Harbor Laboratory Press 0-87969-412-2/92 $3 + 00

centromeric He-T sequences (including representatives from the Y chromosome), and the *Responder* component of segregation distortion. However, probes based on the highly repeated, simple sequence, satellite DNA have yielded no YACs containing long tracts of these sequences.

❑ Approximately 90% of the euchromatic genome is included in the 1193 mapped YACs. There are 149 cytological contigs averaging 638 kb in size; 26 of the cytological contigs exceed 1 megabase (Mb), and the maximum is 4.2 Mb. Cytological contigs are defined based on overlapping regions of hybridization in the salivary gland chromosomes. The definition is conservative in the sense that, in order for a pair of YACs to be considered overlapping, the cytological overlap must include at least two adjacent bands. The cytological contigs are separated by 154 gaps, among which the median size is 50 kb.

❑ The statistics of the Poisson distribution imply that, in mapping strategies based on randomly selected clones, it requires 60% of the total effort to increase from 86.5% genome coverage (two genome equivalents) to 99.3% coverage (five genome equivalents). Therefore, in the next stage of the project, random mapping of YACs gives way to the isolation of bacteriophage P1 clones, which have 75–100-kb inserts, using sequence-tagged-site (STS) probes obtained from previously sequenced genes, nonredundant cDNA clones, and insertion sites of the transposable element *P*. Coverage of the genome with two genome equivalents of P1 contigs averaging 128 kb would require approximately 1700 STS markers.

❑ The P1 contigs obtained by STS screening will close the majority of the gaps in the YAC-based map, creating a nearly continuous physical map, while simultaneously providing convenient substrates for DNA fingerprinting, restriction mapping, or sequenclng, as well as the potential for germ-line transformation. The physical map generated by these methods will be richly annotated with biologically important information that will further the understanding of the biology of *Drosophila* and help to identify and ultimately sequence all the genes in the organism.

INTRODUCTION

The familiar fruit fly, *D. melanogaster*, was among the organisms chosen earliest as a model for studies in genetics. Commensal with humans and spread throughout the world from its native Africa, the species was first described formally by Meigen in 1830 and first intentionally cultivated in the laboratory by the entomologist C.W. Woodworth (Sturtevant 1965). The organism was taken up briefly by the geneticist W. E. Castle in 1901 (Roberts 1986), but its fame and prominence as a model organism came with the work of Thomas Hunt Morgan and collaborators that was carried out in the legendary "fly room" at Columbia University in the years 1909–1928 (Sturtevant 1965; Carlson 1966).

It was also in 1909, for the first and last time in his 24 years at Columbia, that Morgan gave the opening lectures in the undergraduate course in zoology. Among the students were Alfred H. Sturtevant and Calvin B. Bridges. They were both attracted to Morgan and asked to be taken into his laboratory (Sturtevant 1965). There they began their work with *Drosophila*, and within a few years, *Drosophila* had become the premier organism for experiments in animal genetics. Its small size, large number of progeny, short generation time, and ease of culture and manipulation made it ideal for studies of transmission genetics and mutagenesis. In the early 1930s, the giant, exquisitely banded, polytene *Drosophila* chromosomes were discovered in the larval salivary glands, and soon *Drosophila* was also one of the preeminent organisms for cytogenetics (Painter 1933).

Through the years, geneticists have been attracted to *Drosophila* for the rich repertoire of mutations affecting virtually every aspect of the organism's morphology, development, and behavior. Today the opportunities for genetic manipulation and study of *Drosophila* include more than 3,750 described genes, 1,300 mapped transposon insertions, and approximately 18,000 mapped breakpoints in chromosome aberrations (Merriam et al. 1991). The late Larry Sandler once said that, in his opinion, the great advantage of working with *Drosophila* is that any chromosome configuration you can draw you can make. Of course, he was using hyperbole to make a point, but on the other hand, there are surprisingly few constraints on chromosome manipulation other than the need for viability and fertility of the organism and the ability of the chromosomes to undergo replication and segregation.

Supported by the infrastructure of data and resources available for research, *Drosophila* has remained one of the key model organisms for research in molecular biology, cell biology, developmental biology, neurobiology, and evolution, more than 80 years after Morgan first decided to give *Drosophila* a try. Therefore, it was both inevitable and logical to include *Drosophila* among the select few model organisms whose complex genomes were chosen for study under the auspices of the Human Genome Project (Watson and Cook-Deegan 1991). Because of the unique experimental advantages of this organism, the *Drosophila* genome project demonstrates how biology and the technology of complex genome mapping can be joined together in a harmonious union. In this chapter, we summarize the mapping strategy; describe the virtual completion of the initial stage of the project, including the current status of the map; and discuss future directions of the research.

MAPPING STRATEGY USING POLYTENE CHROMOSOMES

The initial experimental strategy for producing a high-resolution physical map of the *Drosophila* genome was based on DNA fragments cloned in YACs, which were assigned to genomic locations by in situ hybridization with the polytene salivary gland chromosomes (Garza et al. 1989a; Ajioka et al. 1991; Hartl et al. 1992).

Choice of vector system

The strategy was based on cloned DNA fragments in order to have direct access to sequences of interest (Merriam et al. 1991), whereas alternative methods for physical mapping have lower resolution than the polytene chromosomes. Among the possible types of clone-based strategies, we initially chose YACs because of the large DNA fragments that can be accommodated (Burke et al. 1987). The initial reliance on YAC clones was justified by the history of efforts in genome mapping using alternative vectors. For example, cosmid mapping, in which overlapping clones forming a "contig" across a contiguous region of the genome are identified by means of restriction-fragment fingerprint patterns, generally yields a large number of contigs averaging under 150 kb (Coulson et al. 1986). Although cosmid contigs are tremendously useful for many purposes, assembling them into a physical map and establishing long-range continuity has required YAC clones to bridge between them (Coulson et al. 1988; Kafatos et al. 1991). When the *Drosophila* project began, cosmids and YACs were the only available choices of large-insert vectors. More recently, a cloning system with vectors based on bacteriophage P1 has been developed, which appears to offer a reasonable compromise between the technical advantages of cosmids and the larger DNA insert size of YACs (Sternberg 1990; Pierce and Sternberg 1991). As discussed below, the P1 system has been assigned a primary role in the post-YAC phase of the *Drosophila* project (Smoller et al. 1991).

Polytene chromosomes

Our approach was also designed to maximize the use of the giant polytene chromosomes of the larval salivary glands. The polytene chromosomes of *D. melanogaster* consist of five chromosome arms, each averaging about 230 μm in length, and one small chromosome arm, about 10 μm in length. The arms are associated in their centromeric regions to form a chromocenter (Lefevre 1976; Ashburner 1989). Each polytene chromosome arm is formed by about a 1000-fold lateral replication of the paired homologous chromosomes, resulting in a bundle of chromatin with a thickness of about 3–5 μm, along which each region has a characteristic pattern of transverse bands. The chromocenter contains the heterochromatic portions of the chromosomes, which are located primarily around the centromeres, plus the entirety of the Y chromosome. Approximately 30% of the genome is heterochromatic and includes a number of highly tandemly repeated, simple sequence, satellite DNA sequences (Lohe and Brutlag 1986). Relative to the euchromatic portions of the chromosomes, which undergo multiple rounds of DNA replication during polytenization, the heterochromatic portions remain underreplicated (Rudkin 1969). On the other hand, since most of the genes in *Drosophila* are located in euchromatin, the salivary gland chromosomes give immediate cytogenetic access to the two-thirds of the genome with the highest density of genetic information.

Detailed analysis of the bands in the salivary gland chromosomes was undertaken by Morgan's student, Calvin B. Bridges (Bridges 1935). The five long chromosome arms are the X (the acrocentric sex chromosome), 2L and 2R (the left and right arms of chromosome 2), and 3L and 3R (the arms of chromosome 3); the short polytene element is the euchromatic portion of chromosome 4. Altogether, Bridges recognized about 5059 bands (Lefevre 1976). This number is in reasonable agreement with the more recent estimate of 5198 based on electron-microscopy (Heino et al. 1992). However, the agreement is partly the coincidence of offsetting overcounts and undercounts. About 500 bands that Bridges designated as doublets, meaning two closely apposed bands that frequently appear as one, are in fact single bands when observed with the electron microscope; in addition, about 600 bands that were not apparent to Bridges using the light microscope can be resolved with the electron microscope (Sorsa 1988).

On the other hand, Bridges's drawings of the chromosome banding patterns (see Fig. 6) include critical landmarks for finding one's place in the polytene chromosomes. Many of the landmarks have colorful and descriptive names—four brothers, shoe buckle, goose neck, Chinese lanterns, road apple, duck's head, and so on (Ashburner 1989). The Bridges drawings, connected to photographic maps illustrating each region as observed in its most typical form, are still the primary frame of reference for the locations of genes, chromosome puffs, rearrangement breakpoints, insertions of transposable elements, cloned DNA fragments, and other cytogenetic features (Lefevre 1976). The Bridges system has stood the test of time because, as Lefevre (1976) noted, "The utility of Bridges' system depends implicitly on the proposition that you and I, without disagreement, can identify in our slides precisely those bands that Bridges so clearly drew and labeled. It is a measure of his genius that, in fact, this implicit assumption is often justified—we can consistently identify particular bands that Bridges drew."

Bridges divided the banded portions of the chromosomes into 102 lettered sections: 1–20 on the X, 21–40 on 2L, 41–60 on 2R, 61–80 on 3L, 81–100 on 3R, and 101–102 on chromosome 4. The tips of the chromosome arms are in sections 1 (X), 21 (2L), 60 (2R), 61 (3L), 100 (3R), and 102(4). The numbered sections of the polytene chromosomes are each divided into six lettered subdivisions, designated A through F, within which the individual bands are numbered sequentially from left to right. The average number of bands per lettered subdivision is 8 or 9, but the range is very wide (2–28). It should be noted here that the bases of the chromosomes, sections 20 (X), 40 (2L), 41 (2R), 80 (3L), 81 (3R), and 101(4), demarcate the euchromatin-heterochromatin junction, which is sometimes designated β-heterochromatin because of its irregular banding pattern; in molecular terms, these junctions are a very long distance away from the α-heterochromatin located in the pericentromeric regions of the chromosomes.

Genome size and present coverage

The size of the *Drosophila* genome is estimated as 165 Mb (Rudkin 1961; Ashburner 1989). Assuming that the euchromatin accounts for two-thirds of the genome, the amount of genomic DNA represented in the polytene arms is 110 Mb. The average polytene band therefore represents a little over 21 kb of genomic DNA, and hence hybridization in situ with a YAC clone containing a 210-kb insert of *Drosophila* DNA would be expected to label 10 bands. The actual average number of bands spanned per YAC is 10.2 (Fig. 1), which is in remarkable agreement with the naive expectation. On the other hand, there is wide variation around this average (the s.d. is ±6.7 bands), in part because the DNA content varies substantially among bands, ranging between the extremes of 5 and 75 kb (Heino et al. 1992). Generally speaking, YACs derived from regions of the genome where the staining intensity of the bands is light tend to hybridize with many bands, whereas those derived from regions where the staining intensity is heavy tend to hybridize with few (Ajioka et al. 1991).

The present summary of the project is based on an analysis of the cytological locations of 1193 euchromatic YAC clones having an average insert size of approximately 210 kb. The total DNA mapped is about 250 Mb, which is approximately 2.3 euchromatic genome equivalents. Assuming that all euchromatic sequences have an equal chance of representation in the YAC libraries, then about 90% of the euchromatic sequences should be included at least once among the mapped clones.

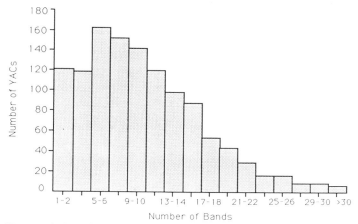

Figure 1 Distribution of number of polytene chromosome bands spanned by hybridization with the *Drosophila* DNA present in each YAC. For example, 121 YACs hybridize with 1–2 bands, 119 with 3–4 bands, 163 with 5–6 bands, and so on. The average number of bands covered per YAC insert is 10.2.

EVALUATION OF THE YAC CLONING SYSTEM

When the *Drosophila* mapping project began, there were reservations about whether the strategy could succeed. Legitimate concerns included the possible instability of YAC clones, the possibility that many genomic sequences might be unclonable or inefficiently cloned in YACs, potential difficulties in screening YAC libraries, the occurrence of chimeric YACs with DNA sequences from noncontiguous genomic regions, and the general inexperience of most laboratories in working with yeast cells and with large DNA molecules. In the past few years, many of these concerns have moderated (Hieter et al. 1990). Data from humans and a variety of model organisms, including *Drosophila*, show that most YAC clones are stable, their inserts are colinear with the genomic sequences from which they derive, and there appears to be no systematic bias in the euchromatic sequences that can be cloned. Screening protocols using colony hybridization have been improved, and methods based on the polymerase chain reaction have been developed that are particularly efficient for screening large libraries (Green and Olson 1990b). Experience in working with large molecules has spread in the scientific community, and strategies and protocols for working with YACs are becoming widely available. On the other hand, despite the technical progress, YACs are still not as easy to work with as bacterial host-vector systems. Furthermore, in human YAC libraries, the very substantial frequencies of chimeric clones, in which two DNA fragments from noncontiguous regions of the genome are joined together, remain a serious technical obstacle to efficient mapping. The frequency of chimeric clones in some human YAC libraries is 50% or more (Green and Olson 1990a), although chimeric clones appear to be much less frequent in the *Drosophila* YAC libraries.

Preparation of YAC libraries containing *Drosophila* DNA

To apply the YAC cloning technology to the *Drosophila* genome, we constructed several different types of YAC libraries containing large inserts of *Drosophila* DNA (Garza et al. 1989a; Danilevskaya et al. 1990; Ajioka et al. 1991). The YAC libraries are summarized in Table 1. They include a total of 7488 clones arrayed in 78 microtiter plates, comprising about 10 genome equivalents, of which 1193 have been assigned to positions in the salivary gland chromosomes.

Random sheared DNA   The initial library of 768 clones was made using the YAC vector pYACP-1 (Garza et al. 1989a), which is a modified version of the prototype vector pYAC2 (Burke et al. 1987). In addition to the functional elements necessary for the selection and maintenance of clones in yeast, the vector contains the ends of the *P* element (Bingham et al. 1981) and the *hsp70:G418* resistance gene (Steller and Pirrotta 1985). To prepare the library, high-molecular-weight source DNA from an Oregon RC strain was isolated from lysed nuclei and purified by CsCl density gradient centrifugation (Bingham et al. 1981). This procedure yielded DNA molecules up to approximately 500 kb in length. The DNA ends were repaired with bacteriophage T4 DNA polymerase, and fragments were separated by size in sucrose velocity gradients (Carle and Olson 1984). Collected fractions were assayed by field inversion gel electrophoresis, and those fractions with DNA fragments larger than 120 kb were pooled and concentrated in a collodion bag. Size-fractionated DNA was used as insert DNA for ligation with prepared pYACP-1 vector arms, size-fractionated and concentrated again, and then used to transform yeast strain AB1380 (Burke et al. 1987).

Table 1 Composition of *Drosophila* YAC libraries

YAC vector	Type of insert	Designation	Total clones	Mapped clones
pYACP-1	randomly sheared	DY	768	272
pYAC5	*Not*I partial	N	2688	502
pYAC5	*Eco*RI partial	E or R	4032	419
Total			7488	1193

Not*I* *library* This library consists of 2688 clones. Gastrula cells from Oregon RC were embedded in plugs of 1% low-melting-temperature agarose and partially digested in situ with *Not*I. The plugs were transferred to tubes, vector arms from the vector pYAC5 (Burke et al. 1987) were added in an equal mass as genomic DNA, and the mixture was incubated to melt the agarose. Ligation buffer and T4 DNA ligase was added to each tube, and incubation was carried out to attach genomic fragments to the vector arms. The methods are described in detail by Danilevskaya et al. (1990).

Eco*RI* *library* A total of 4032 YAC clones were isolated that contain inserts of *Eco*RI fragments obtained by partial digestion of genomic DNA from an isogenic strain (iso-1) of genotype *y; cn bw sp*. The *Eco*RI fragments were prepared from genomic DNA essentially as described by Garza et al. (1989a), except that the high-molecular-weight DNA was very lightly digested with *Eco*RI prior to size fractionation to enrich for the size range 150–550 kb. The resulting fragments were ligated onto the vector arms of pYAC4 (Burke et al. 1987), and transformation was carried out.

Representation and stability of genomic sequences in YAC clones

We have carried out a number of tests to determine the completeness of representation of different genomic sequences among the YAC clones and to determine whether clones were regularly undergoing rearrangement in yeast. Completeness of coverage was judged on the basis of results of colony hybridizations with both repetitive and single-copy DNA sequences. In some cases, we have been able to compare the pattern of genomic restriction fragments with the pattern obtained from YACs containing the same sequences, and we have generally found them to be in excellent agreement.

Repetitive sequences One of the toughest challenges for any cloning system is the stable cloning of repetitive DNA. Since part of our objective has been to evaluate the adequacy and reliability of the large-molecule cloning systems, we deliberately subjected the systems to a number of tests involving repetitive DNA. The overall results can be summarized in terms of the following types of repetitive sequences:

1. *Ribosomal DNA (rDNA)*. We isolated approximately 50 YAC clones containing rDNA sequences, confirmed them by Southern blotting, and analyzed them for the presence of Type

I and Type II rDNA inserts (Garza et al. 1989a). These clones hybridized to the chromocenter in the salivary gland nuclei. Although the majority of the rDNA-containing YACs were stable, 5–10% of the clones were unstable in the sense of showing multiple YAC bands in one or more of the cultures derived from single yeast colony replicates.

2. *He-T sequences.* This family of repetitive DNA sequences is located in heterochromatin and near the telomeres (Traverse and Pardue 1989). In screening with a He-T clone designated Dm665, originally isolated because of its capacity for autonomous replication in yeast, we obtained approximately 20 clones from the *Not*I library that were confirmed with Southern blots (Danilevskaya et al. 1990). A number of these showed a restriction fragment homologous to Dm665 that is specific to the Y chromosome, and we were able to demonstrate that restriction fragments in genomic digests of females bearing particular portions of the Y chromosome from a series of X-Y translocations coincided with those from YACs containing the homologous sequences (Danilevskaya et al. 1990). Thus, the YAC libraries contain sequences from the Y chromosome, and the clones appear to be stable. More recently, we have identified a series of Y-chromosome YACs containing the *Suppressor of Stellate* locus (Livak 1990), which shares regions of homology with Dm665 and is located in many of the same YAC inserts (A.R. Lohe and D.L. Hartl, unpubl.).

3. *Responder (Rsp).* The *Responder* component of the segregation distorter system is a repetitive DNA sequence located in the heterochromatin of 2R, the number of copies of which determines resistance to segregation distortion (Wu et al. 1988). We have isolated a *Not*I clone that contains most of the *Rsp* sequences present in the Oregon RC genome. Analysis of this clone demonstrates concordance between the genomic and YAC restriction fragments containing homology with *Rsp*, as well as indicating that *Rsp* sequences occur in clusters separated by stretches of non-*Rsp* DNA (D.A. Smoller and D.L. Hartl, unpubl.). The *Rsp*-bearing YAC hybridizes to the chromocenter of salivary gland nuclei.

4. *Simple satellite sequences.* A number of highly repeated, simple sequence, satellite sequences located primarily in the pericentromeric heterochromatin and the Y chromosome have been analyzed by Lohe and Brutlag (1986), and we have used some of the satellite probes for screening the YAC libraries. The main result is that YAC clones containing long tracts of simple satellite DNA have not been recovered. However, the absence of such clones may reveal more about the physical state

of satellite DNA than it does about the YAC cloning system. On the basis of their experience in trying to clone satellite sequences from *Drosophila* using plasmid vectors in *Escherichia coli*, Lohe and Brutlag (1986) have suggested that the simple satellite sequences may be lost during DNA purification or may prevent efficient transformation of the host, owing to the physical state of the sequences in the genome (e.g., tightly bound, or even covalently attached, to certain proteins). In any case, their underrepresentation in *E. coli* libraries is not merely a matter of difficulty in being propagated, although clone instability may be contributory. The Lohe and Brutlag hypothesis would also explain why YAC libraries lack clones with long tracts of simple sequence satellite DNA, whereas clones containing more complex types of repetitive sequences do occur (e.g., rDNA, He-T sequences, *Rsp*). On the other hand, whether the complex repetitive sequences are present in YAC libraries in proportion to their abundance in the genome is not known.

Single-copy sequences Initial tests of the YAC cloning system were carried out on clones detected with single-copy probes. These included probes for the genes *white, rosy, alcohol dehydrogenase*, and *abdominal-A*. Each probe identified at least one clone that was subsequently confirmed by DNA transfer hybridization of field inversion electrophoresis gels using the appropriate probes, and by direct in situ hybridization with polytene chromosomes. The internal structures of the YAC clones containing *white, rosy*, and *alcohol dehydrogenase* were examined by comparing the lengths of restriction fragments from the YAC clones with those in the Oregon R genome. In each case, the bands from the Oregon R genome were of the expected size and matched equivalent bands from the corresponding YAC clone (Garza et al. 1989a). A detailed analysis of 160 kb of cloned DNA derived from the Bithorax complex yielded similar results (D. Garza and D.L. Hartl, unpubl.), as did a series of clones from the Antennapedia complex (Garza et al. 1989b; Ajioka et al. 1991). In addition, each of the euchromatic YAC clones yielded a single artificial chromosome of the expected size upon repeated subcloning.

In summary, our experimental tests of the YAC cloning system imply that most euchromatic YAC clones contain faithful replicas of DNA sequences present in the *Drosophila* genome and that they are genetically stable in the sense that they do not appear to rearrange at frequencies high enough to be detected during routine subcloning in yeast. Nevertheless, detailed tests have been carried out on a small scale with a relatively small number of clones, and we cannot exclude the possibility that some fraction of the euchromatic YAC clones may be rearranged or unstable.

YAC-BASED *DROSOPHILA* GENOME MAP

Although some targeted screening of the YAC libraries has been carried out for purposes of system evaluation (Garza et al. 1989a,b; Ajioka et al. 1990, 1991; Ochman et al. 1990; Smoller et al. 1991), most of the genome mapping has thus far been carried out with YAC clones selected at random from the libraries. The present YAC-based *Drosophila* genome map is based on the cytological locations of 1193 euchromatic YAC clones having an average insert size of approximately 210 kb, and it includes about 90% of the euchromatic sequences (Garza et al. 1989a; Ochman et al. 1990; Ajioka et al. 1991; Smoller et al. 1991; Hartl and Palazzolo 1992; Hartl et al. 1992). For in situ hybridization, bands containing individual YAC chromosomes were excised from field inversion gels, the DNA was purified using the sodium iodide–glass powder method (Vogelstein and Gillespie 1979) and labeled with biotin-dCTP by the random hexamer method (Feinberg and Vogelstein 1984), and in situ hybridizations to polytene chromosome preparations were carried out according to the method of Langer-Safer et al. (1982).

Table 1 gives the distribution of mapped clones derived from the YAC libraries containing DNA fragments with random ends, *Not*I ends, or *Eco*RI ends. The size distribution of the clones is shown in Figure 2. Each YAC contains, on the average, roughly the same quantity of DNA as the average lettered subdivision in the salivary gland chromosomes (Sorsa 1988). Unfortunately, figures of conventional size are too small to illustrate the positions of the YACs on the chromosome map with an acceptable degree of resolution. However, the October 11, 1991, issue of *Science* (**254:** 165–344) includes a 55-cm x 80-cm fold-out poster with a drawing of the salivary gland chromosomes, which also shows the positions of the YACs along with cosmid contigs, lengthy chromosome walks, many genes, and much other useful information (Merriam et al. 1991).

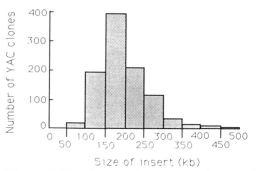

Figure 2 Distribution of insert sizes of *Drosophila* DNA included among the mapped YACs.

Examples of the density of coverage are illustrated in Figures 3 and 4. Figure 3 depicts coverage of the distal half of the X chromosome (sections 1–10), Figure 4 the distal half of 3R (sections 91–100). The banding patterns appear very unfamiliar because the widths of the bands have been scaled in proportion to their estimated DNA content, determined from electronmicroscopy (Heino et al. 1992). The extent of hybridization with each YAC is indicated by a horizontal line segment beneath the relevant bands, with the line arbitrarily extending halfway into the bands in which the YAC hybridization terminates. The single bands recognized by Bridges (1935) are indicated by the short vertical tick marks, and the single bands he considered doublets are indicated by Y-shaped ticks. In some cases, electronmicroscopy indicates that there are additional bands (Heino et al. 1992), which are indicated as bands without any ticks.

It may appear from Figures 3 and 4 that the distal half of the X is covered with YACs more sparsely than the distal half of 3R, and this is

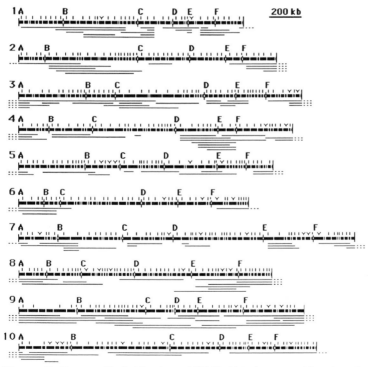

Figure 3 Extent of hybridization of YACs (thin horizontal lines) in the distal half of the X chromosome (polytene chromosome sections 1–10). The bands (thick horizontal lines) have been scaled according to the estimated quantity of DNA that each band contains.

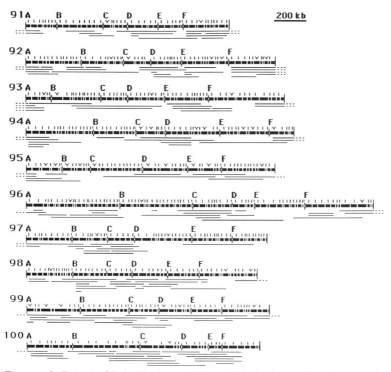

Figure 4 Extent of hybridization of YACs (thin horizontal lines) in the distal half of the right arm of chromosome 3 (polytene chromosome sections 91–100). The bands (thick horizontal lines) have been scaled according to the estimated quantity of DNA that each contains.

actually the case. The estimated DNA content in sections 1–10 is 13,351 kb, and that of sections 91–100 is 12,288 kb (Heino et al. 1992). Altogether, there are 261 YACs in these sections, and if they were distributed in proportion to the DNA content, then 135.7 YACs would be expected in sections 1–10 and 125.3 in sections 91–100. The observed numbers are 108 and 153, respectively, and the discrepancy is statistically highly significant ($\chi^2 = 11.8$, $p < 0.001$). This kind of discrepancy occurs when the X chromosome is compared with any of the autosomes (Ajioka et al. 1991; Hartl et al. 1992), but it is only an apparent discrepancy. Since the YAC libraries were constructed from approximately equal numbers of males and females, the DNA preparations contained three X chromosomes and one Y chromosome for each four sets of autosomes; hence, the representation of the X chromosome among the YACs, relative to that of autosomes, should be 3/4. Taking this correction into account for the data in sections 1–10 versus 91–100, the expected numbers of YACs are 117.5 and 143.5, respectively, and now the agreement with the observed numbers is very good ($\chi^2 = 1.4$, $p = 0.25$).

Cytological contigs

One of the unique advantages of *Drosophila* for genome studies is that continuity of coverage can be assessed cytologically in the polytene chromosomes. As noted, the average mapped YAC contains an insert of 210 kb and hybridizes with an average of about 10 salivary chromosome bands. Consequently, overlaps between YAC clones can often be identified cytologically. To compare the current status of the *Drosophila* physical map with that of other organisms, it is convenient to define a *cytological contig* as an uninterrupted sequence of salivary chromosome bands and interbands present in at least one YAC. To avoid confusion, it should be emphasized that cytological contigs are determined at the level of resolution of the light microscope in the polytene salivary gland chromosomes, not using restriction-fragment fingerprints, restriction maps, end-specific probes, or other molecular criteria.

In identifying cytological contigs based on YACs, we use the rather conservative rule that, in order for genomic coverage between a pair of adjacent YACs to be considered continuous, the cytological overlap between the YACs must include at least two adjacent bands; since, even in the minimal case of a two-band overlap, both YACs must include the interband between the bands, and each of the bands itself must be completely included in at least one of the YACs. Application of this rule is illustrated in Figure 5 for some YACs in subdivision 1B of the X chromosome. In this case, the bands are numbered according to Bridges (1935), and extra bands appearing in the electron microscope are denoted + (Heino et al. 1992). For purposes of defining a cytological contig, the YACs E00-52 and DY174 would be regarded as nonoverlapping; although they both extend into 1B3, and may, in fact, overlap at the molecular level, the overlap, if it exists, is not apparent cytologically. (However, in this particular example, it is clear that any gap between E00-52 and DY174 must be included in N13-23.) Over in the region 1B11-1B12, the YACs N13-23 and N23-03B would be regarded as over-

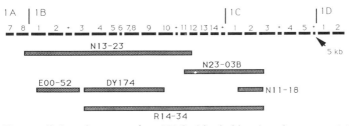

Figure 5 Localizations of six YACs (shaded bars) in the region 1A8-1C3 near the tip of the X chromosome. For purposes of defining cytological contigs, YACs like E00-52 and DY174, which terminate in the same band, are regarded as nonoverlapping. A minimal overlap of two bands is required for YACs to be regarded as overlapping, for example, N13-23 and N23-03B.

lapping: They must overlap, since N13-23 extends through 1B11 and the interband into 1B12, and N23-03B extends through 1B12 and the same interband into 1B11.

It should be emphasized that the definition of a cytological contig depends on the relatively high cytological resolution of the polytene salivary gland chromosomes, since the bands average only about 10% of the size of the average YAC insert. Metaphase chromosomes from *Drosophila* or any other organism would not allow this level of resolution. On the other hand, a note of caveat emptor is also in order. Even using the definition of a cytological contig that requires a minimal two-band overlap, there are two principal inherent limitations in the method. First, whereas the cytological localizations of the YACs are generally accurate, the specification of the particular bands that mark the end points is based on visual inspection that inevitably includes a degree of subjectivity depending to a significant extent on the quality of the banding in the relevant region and also on the degree of labeling and the efficiency of hybridization. Second, the accuracy of the estimates of DNA content per band is uncertain. These estimates are based on the thickness and density of the bands as observed in the electron microscope in thin sections (Heino et al. 1992), and they may be in error to the extent that the bands are nonuniform in different planes of section or in different genomic regions.

Hence, the analysis of cytological contigs should be regarded as approximate, and having as its main purpose a summary of the *Drosophila* physical map at its present stage of development in terms that allow comparison with genome projects in other organisms. The ultimate proof of clone overlap is the fact that they share DNA sequences, and some methods for establishing molecular contigs are discussed later in this chapter.

With these cautionary notes as background, the distribution of cytological contigs in the euchromatic genome of *Drosophila* is illustrated in Figure 6. Above the chromosome drawings, the numbered sections are indicated, and the lettered subdivisions A–F are separated by tick marks. The cytological contigs are shown beneath the chromosomes, and the sizes are indicated in megabases for those greater than 0.5 Mb.

The size distribution of the cytological contigs is shown in Figure 7. For cytological contigs smaller than 1 Mb, the scale is in kilobases, and for those larger than 1 Mb the scale is in megabases. The size of each cytological contig has been estimated from the data of Heino et al. (1992) by summing together the DNA content of each band completely included within the contig plus one-half of the DNA content of the bands in which the contig terminates. The DNA content of the interbands was not incorporated explicitly, since the estimated DNA content per interband averages less than 1 kb (Beermann 1972). However, the estimated

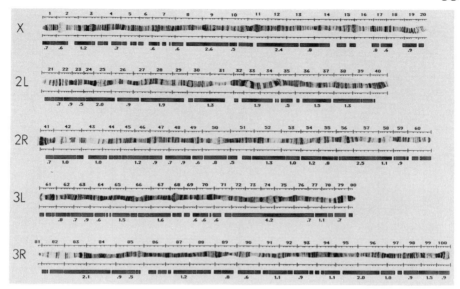

Figure 6 Map of 149 cytological contigs in the *Drosophila* euchromatin, indicating the size in megabases of all cytological contigs greater than 0.5 Mb.

DNA content per band of Heino et al. (1992) has been normalized to a total euchromatic genome size of 110 Mb, and so the interbands are implicitly taken into account. Altogether, Figure 7 includes 149 cytological contigs, 26 of which exceed 1 Mb in size. The average size is 638 kb, and the maximum size is 4.2 Mb (in sections 72–76). Considering the euchromatic genome as a whole, approximately 86.5% of the DNA sequences are included in cytological contigs.

The actual number of contigs in the 1193 YACs is probably smaller than 149, since situations like the YACs E00-52 and DY174 in Figure 5 are considered as nonoverlapping, when, in fact, they may overlap at the molecular level. Some insight into the actual number of contigs can be gained from the results of computer simulations, assuming 1193 YACs of 210 kb size randomly distributed in a genome of 110 Mb (Palazzolo et al. 1991). In this case, the actual number of contigs is 122 (average of 10 simulations, with s.e.m. ±2), which suggests that about 27 of the gaps in Figure 6 would actually be closed if appropriate molecular studies were carried out. The observed number of 149 cytological contigs matches that obtained from simulations with 1193 YACs of size 190 kb, which implies that the effective length of the YACs is reduced by about 20 kb by the requirement for a minimal two-band overlap for YACs to be considered in the same cytological contig. The 20-kb difference is about the same as the level of resolution of in situ hybridization (Merriam et al. 1991), and it reflects one of the limitations of cytological analysis.

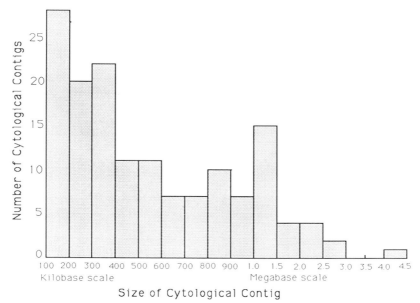

Figure 7 Distribution of sizes of 149 cytological contigs. The range from 0.1 Mb to −1.0 Mb has been expanded and is scaled in units of 100 kb. The mean size of the cytological contigs is 638 kb, and there are 26 cytological contigs greater than 1 Mb.

Depth of coverage

Most of the polytene chromosome bands that are included in the mapped YACs are actually included in more than one YAC. This aspect of the data is shown in Figure 8, which shows the depth of coverage in terms of the number of times each salivary chromosome band has been cloned independently into different YACs. Including the class of bands not present in any YAC (550 bands out of 5157 in the major chromosome arms), the average depth of coverage is 2.36, which means that an average band is included in 2.36 YACs. Therefore, assuming Poisson coverage, the expected proportion of the genome covered in the mapped YACs is 1 − exp(2.36) = 90.6%. Furthermore, among the bands present in one or more YACs, the depth of coverage averages 2.65, which means that approximately 75% of the bands that have been cloned are actually present in more than one YAC.

Location and size distribution of gaps

In mapping strategies based on contig assembly, the benchmarks of progress are usually the number of contigs and their sizes (Coulson et al. 1991). Without polytene chromosomes, or some alternative high-

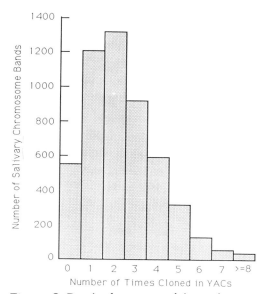

Figure 8 Depth of coverage of the euchromatic genome in terms of the number of times that each polytene chromosome band has been cloned in independent YACs. About 89% of the bands are included in at least one YAC, and among these an average of 2.65 YACs hybridizes with each band.

resolution cytogenetic assay, little or nothing can be determined about the spatial relations between the contigs or the size distribution of the gaps. In contrast, use of the *Drosophila* polytene chromosomes as a guide in genome mapping allows gaps to be determined as easily as the coverage, and this is one of the uniquely advantageous features of this organism.

The gaps in coverage of the euchromatin in the major chromosome arms are summarized in Figure 9. For purposes of defining gaps based on cytological criteria, two YAC clones that terminate in the same polytene chromosome band are considered nonoverlapping, and hence a small gap is assumed to occur between them, even though, at the molecular level, the YACs may overlap. For each cytological gap, the size was estimated by summing together the DNA content of each band located completely inside the gap plus one-half of the DNA content of the bands included in YACs at the ends of the gap.

Altogether, there are 154 cytological gaps. The mean size is 90 kb, but the median size is only 50 kb because the distribution is skewed to the left. The maximum cytological gap is about 730 kb, which spans region 6C2–6F1,2 on the X chromosome. (The terminology 6F1,2 refers to a band that appears unitary in the electron microscope but which Bridges [1935] considered as a doublet.) The next largest gaps are about 500 kb, spanning 85F2–86B3 in the right arm of chromosome 3; then

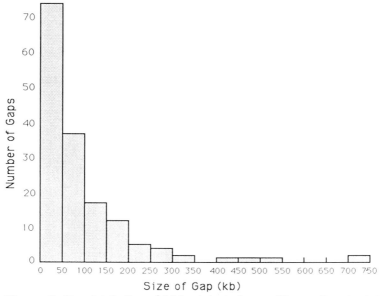

Figure 9 Size distribution of 154 cytological gaps. The median size is 53 kb and the average size is 90.5 kb.

450 kb, on the X chromosome spanning 18C5–18E5; and 430 kb, in the left arm of chromosome 2 spanning 31E1–32A1.

Cytological gaps are defined conservatively in the sense that two YAC clones terminating in the same band are assumed not to overlap, and so the estimate of 154 as the number of gaps in the YAC map of the major euchromatic chromosome arms is probably a maximum. As noted in the section on cytological contigs, computer simulations suggest that the true number of contigs is probably around 122; the same simulations imply that the true number of gaps is around 127. Moreover, considering the 154 cytological gaps as defined, 71% are smaller than 100 kb, and 62% are smaller than 80 kb (Fig. 9).

Underrepresentation of chromosome 4

Thus far, we have made no explicit mention of YACs that hybridize with chromosome 4. This tiny chromosome contains an estimated 1.3 Mb of euchromatic DNA, or about 1.2% of the euchromatic genome. Hence, with random euchromatic representation among 1193 YAC clones, about 14 would be expected to hybridize with chromosome 4; however, only 3 were observed, and the deficiency is statistically highly significant (χ^2 = 8.7, p = 0.005). The reason for the discrepancy is unknown. Although chromosome 4 may contain more than its share of moderately repetitive DNA (Miklos et al. 1988), the euchromatic portion appears to become polytene like the other autosomes. In any event, if the four- to fivefold underrepresentation of chromosome 4 in the YAC libraries is not a peculiar feature of *Drosophila*, but rather reflects some general characteristic of the organization or content of the DNA sequences in chromosome 4, then the underrepresentation of this chromosome is potentially quite troublesome for mapping projects in other organisms that depend exclusively on YACs. At the very least, mathematical analysis and computer simulations of the efficiency of various mapping strategies may be misleading (Barillot et al. 1991; Palazzolo et al. 1991), since one of the main assumptions is that the YAC libraries are equally representative of all relevant portions of the genome.

STRATEGY FOR CLOSURE

Closure, or the approach to completion, of the *Drosophila* map based on YACs has reached a stage where a change in strategy is in order. Mapping strategies based on randomly selected clones, including the strategy of in situ localization, are severely limited, after about two genome equivalents have been mapped, by certain statistical properties of the Poisson distribution (quite apart from possible nonrandomness in the cloning efficiency of different genomic regions) (Palazzolo et al. 1991). The nature of the Poisson limitation can be illustrated as follows: Since $1 - \exp(-2) = 0.865$, whereas $1 - \exp(-5) = 0.993$, mapping two genome equivalents gives 86.5% coverage, and mapping five genome equivalents gives about 99.3% coverage. Hence, with random clone selection, it requires the mapping of three additional genome equivalents to increase coverage from 86.5% to 99.3%; or, in other words, it requires 60% of the total effort and expense to complete the final 12.8% of the map. This comparison discourages continuing the random selection of clones at a stage when approximately two genome equivalents have already been mapped, and alternative nonrandom clone selection strategies can significantly improve progress and decrease labor and expense (Palazzolo et al. 1991).

Advantages of the bacteriophage P1 cloning system

For the next phase of the *Drosophila* project, we have chosen to rely on bacteriophage P1 clones instead of YACs. The P1 vector contains a packaging site (*pac*) that allows approximately 115 kb to be packaged into infectious particles able to infect *E. coli*, a pair of *loxP* recombination sites at which the Cre recombinase circularizes the DNA inside the host cells, and a bacterial kanamycin-resistance marker used for selection (Sternberg 1990; Pierce and Sternberg 1992). P1 vectors have also been developed that contain a *sacB* gene from *Bacillus* having a unique cloning site between the promoter and the coding sequence, which allows selection of vectors containing inserted DNA, since the insert prevents production of the *sacB* gene product (levansucrase) and allows *E. coli* cells to survive in 2% sucrose (Pierce et al. 1992). In the P1 system, cloned DNA is normally maintained at one copy per cell by the plasmid replicon, but there is a P1 lytic replicon under the control of a *lac* promoter than can be induced to give amplification to 20–30 copies per cell when desired (Sternberg 1990). Typically, P1 clones containing inserts in the size range 75–100 kb are obtained at an efficiency of greater than 10^4 clones per microgram of source DNA, which is 10–100 times more efficient than the recovery of YACs (Sternberg 1990).

Some of the arguments for the use of P1 clones in genome analysis are as follows. First, P1 clones have a number of technical advantages not found in YACs, primarily emanating from the fact that the P1 vector system has *E. coli* as the host and that large quantities of high-quality DNA can be obtained. Hence, the method of colony filter hybridization is quite reliable for screening P1 libraries, and P1 clones can provide ample material for technical manipulations such as DNA fingerprinting, restriction mapping, or sequencing, as well as for such potential biological applications as germ-line transformation (although the feasibility of germ-line transformation of *Drosophila* with P1 clones has yet to be demonstrated). Second, although the insert size of P1 clones is not as large as that of YACs, the average is about 80 kb, and hence the inserts in P1 clones are large enough to provide valuable cytogenetic information when mapped by in situ hybridization. (The average P1 clone would be expected to hybridize with 3–4 bands in the salivary gland chromosomes.) Third, the majority of the gaps remaining in the YAC-based map of the euchromatin are smaller than the insert size in P1 clones. As noted in connection with distribution of cytological gaps, 71% of the gaps are smaller than 100 kb and 62% are smaller than 80 kb. Since 80 kb is approximately the average insert size among bacteriophage P1 clones (Smoller et al. 1991), and 100 kb is near the maximum P1 insert size (Sternberg 1990), most of the remaining gaps can be bridged by P1 clones.

Mapping with sequence-tagged sites

Olson et al. (1989) have defined a sequence-tagged site (STS) as a site in the genome uniquely determined by its DNA sequence. There are a number of advantages to using STS markers as the primary landmarks in genome mapping. First, the sequences defining random STS markers are relatively simple to generate, and assays using the polymerase chain reaction with appropriate synthetic oligonucleotide primers can be used to identify clones containing any STS. Second, STS markers can be used to confirm overlaps between clones, since two nonchimeric clones containing the same STS must overlap. Third, the definition of STS markers in terms of unique DNA sequences frees the genome map from any particular collection of clones, since any trained scientist should be able to use the defining sequence to identify the presence of the STS in any DNA sample or library of interest. An additional advantage of STS markers in *Drosophila* is that they can be mapped cytogenetically by in situ hybridization, which not only allows progress to be assessed as the mapping proceeds, but also serves to detect any chimeric clones, misplaced clones, or other potential errors.

Biological choices of STS markers

In its original form, the proposal to emphasize STS markers in genome mapping envisioned the sites as being chosen completely at random from anonymous DNA sequences (Olson et al. 1989). More recently, a number of groups have emphasized the potential advantages of using biologically meaningful sequences as STS markers (Adams et al. 1991; Hartl et al. 1992). An attractive feature of this strategy is that the physical map assembled by means of the STSs becomes annotated with important biological data, and valuable research resources are generated at the earliest possible phase of construction and throughout the development of the map.

Known genes One important class of STS markers in *Drosophila* consists of the more than 500 genes that have been cloned and sequenced (Merriam et al. 1991), since most of these genes have been mapped and thus provide the anchor points for correlating the genetic linkage map with the physical map. Genetic linkage maps have been an important framework for organizing genetic information ever since the first linkage map of *Drosophila* was developed by A.H. Sturtevant (1913). The main limitation of genetic linkage maps is that they are based solely on recombination data and therefore provide no direct physical link or access to the underlying DNA. In contrast, clone-based physical maps depict the relative locations of certain landmarks present in genomic DNA, such as the locations of STS markers or restriction sites. The principles underlying linkage maps and physical maps are quite distinct because the assembly of the two types of maps is operationally very different. Derivation of a linkage map is essentially an extension of pedigree analysis; hence, linkage maps are based on genes as their fundamental components, and the distances are derived from recombination frequencies. Construction of a physical map is essentially a molecular analysis; hence physical maps usually rely on clones as their fundamental components, and the distances between molecular landmarks are ideally measured in nucleotide pairs. Only by assigning genes in the linkage map to locations in the physical map can the two types of representation be correlated.

P-element insertions Another important class of STS markers in *Drosophila* consists of sites defined by mutability via insertion of the transposable element *P* (Engels 1989). A set of approximately 1000 autosomal *P*-element insertions that are lethal when homozygous have been generated (Cooley et al. 1988a,b). Since the insertions are recessive lethals, the mutations identify genes that are essential to survival. The particular *P* element used in the mutagenesis contains a bacterial origin of replication and selectable marker, which enables the genomic sequences flanking the insertions to be isolated by "plasmid rescue," in which the *Drosophila* DNA is extracted, digested with appropriate restriction enzymes, and self-ligated to form circles; the circles are transformed into bacteria in order to select the reconstituted plasmid (Mlodzik et al. 1990). Once cloned in a bacterial plasmid, the genomic flanking DNA can be used as probes for in situ hybridization, library screening, or DNA sequencing to define STS markers.

cDNA sequences A third important source of biologically relevant STS markers consists of coding sequences (Adams et al. 1991; Hartl et al. 1992). Advances in cDNA technology have made it possible to isolate and sort large numbers of cDNA molecules that are corrected for the differential abundance of mRNA sequences found in most cDNA libraries (Palazzolo et al. 1989). The sorting procedure results in a set of nonredundant cDNA sequences that includes representatives of many of the least abundant types of mRNA sequences in the target tissue. A set of approximately 750 sorted cDNA clones has been obtained from transcripts in the adult head of *Drosophila* that are not detectable in the early embryo (Palazzolo et at. 1989). Used as STS markers in the genome map, these cDNA sequences represent important materials for building a database of the molecular components of various tissues in *Drosophila*, starting with the nervous system.

Using multiple sources of STS markers provides a greater likelihood of widespread distribution throughout the genome, since possible biases in the distribution of one set of STS markers may be compensated by markers from another source. For example, because of the way in which they were selected, most of the *P*-element lethals are autosomal. STS markers for the X chromosome must come from known genes or cDNAs.

Building an integrated map

The first step in building an integrated map of *Drosophila* is the use of the STS markers as probes to isolate multiple P1 clones. Each set of P1 clones identified by a particular STS forms a P1 contig, since they all contain the same STS. Since the positions of the mapped genes and *P*-element insertions have already been located in the salivary gland chromosomes, the cytogenetic locations of the P1 contigs can immediately be inferred. The cytogenetic locations of the P1 contigs derived from the sorted cDNAs can be determined directly by in situ hybridization.

In screening a P1 library containing five genome equivalents in which the average insert size is 80 kb, the average P1 contig generated by an STS is approximately 1.6 times the insert size, or 128 kb (Palazzolo et al. 1991). Therefore, the recovery of one genome equivalent of P1 contigs from a 110-Mb genome would require approximately 850 STS markers. Coverage representing two genome equivalents would require approximately 1700 STS markers, and three genome equivalents would require 2550 STS markers. Although 1700 is not a trivial number of P1 library screens, neither is it especially intimidating at a time when filter replicas can be produced in large numbers by robotics (Lehrach et al. 1990). Furthermore, the P1 contigs obtained from the STS markers represent a set of working clones centered on biologically important sites that can be used immediately for DNA sequencing as well as other purposes.

Once the stage is reached when many P1 contigs have been identified and located in the genome, the next step is to obtain a continuous genome map by correlating the P1 contigs with the YAC clones already mapped. In principle, the correlation between the YAC map and the P1 map could be carried out concurrently by screening both the P1 and YAC libraries with the STS markers. However, this strategy lacks efficiency because of the greatly increased screening effort, and it is better to utilize the YACs to connect established P1 contigs to yield a continuous physical map. The major effort along these lines remains to be carried out, but the results of a preliminary feasibility test are summarized in Figure 10, which illustrates the YAC clones in section 67 of the salivary gland chromosomes and the P1 contigs obtained by screening with three eye-specific cDNAs (D.L. Hartl and D. Petrov, unpubl.).

The eye-specific cDNA clones 2g8, 3f12, and 8h7 were used as STS markers to screen a small P1 library (Smoller et al. 1991), and each STS yielded one or more P1 clones that were assigned positions in the salivary chromosome map by in situ hybridization. To determine which YACs contained sequences homologous to the cDNAs, the YACs in the corresponding regions were screened by DNA transfer hybridization using the cDNAs as probes. In the region 67A-67B, two YAC clones hybridized with 2g8 only, one YAC with 3f12 only, and one YAC with both 2g8 and 3f12. Thus, the cDNA hybridizations confirm, at the molecular level, a YAC contig of approximately 600 kb covering much of the region 66F-67B. Similarly, screening with the cDNA clone 8h7 yielded a pair of P1 clones that mapped to the region 67F4-68A1, and 8h7 hybridized with two YACs in this region, confirming a contig of about 420 kb. Between the two contigs in Figure 10 is another approxi-

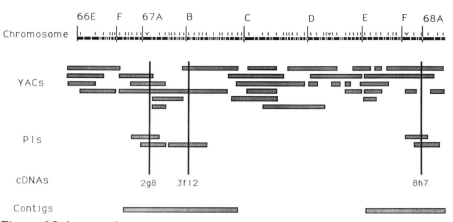

Figure 10 Integrated mapping strategy that uses biologically important STS markers (in this example, the eye-specific cDNAs 2g8, 3f12, and 8h7) to identify a set of working P1 clones and to confirm overlaps between YAC clones at the molecular level.

mately 600 kb that is well covered with YACs that will probably be connected as additional P1 clones are incorporated into the map. The total contig in this region would then be about 1.6 Mb.

CONCLUSIONS

A physical map of *D. melanogaster* covering 90% of the euchromatic genome has been generated by determining the positions of 1193 YAC clones in the polytene salivary gland chromosomes by means of in situ hybridization. The YAC clones have an average insert size of approximately 210 kb, and hence the total DNA mapped is about 250 Mb, or approximately 2.3 euchromatic genome equivalents. Statistical tests suggest that, although chromosome 4 appears to be underrepresented, coverage of the rest of the euchromatic genome with YACs appears to be nearly random. Because of the high resolution possible using the salivary gland chromosomes, cytological contigs can be defined based on cytologically observed overlaps between YACs. For this purpose, two YACs are regarded as overlapping only if they share two or more contiguous bands of hybridization. With this definition, the major chromosome arms are present in 149 cytological contigs, 26 of which are greater than 1 Mb in size. The average size is 638 kb and the maximum size is 4.2 Mb. The cytological contigs are separated by 154 gaps of mean size 90 kb. The median gap size is only 50 kb because the size distribution is skewed.

On the other hand, the degree of YAC coverage of the non-polytenized portion of the *Drosophila* genome, which includes large amounts of heterochromatin flanking the centromeres and the entirety of the Y chromosome, is still unknown. Some heterochromatic sequences are present in the YAC libraries, including complex repeated sequences such as ribosomal DNA, the telomeric and centromeric He-T sequences (including representatives from the Y chromosome), and the *Responder* component of segregation distortion. However, YACs containing long tracts of highly repeated, simple sequence, satellite DNA have not been recovered, possibly because these sequences have a special physical state that militates against their recovery (Lohe and Brutlag 1986).

With 90% coverage, the YAC-based phase of our strategy of mapping the genome of *Drosophila* is complete. The next phase of the project gives primacy to bacteriophage P1 clones because of their size and technical advantages, and it employs a variety of sequence-tagged sites, including known sequenced genes, transposable *P*-element insertions, and cDNA clones, in order to isolate a large number of P1 contigs containing biologically important sequences that can be connected by the YACs into a small number of very large contigs. The resulting genome map and research materials (YAC clones, P1 contigs, biologically significant STS markers) will open up new opportunities for *Drosophila* research, including ready access to the DNA in any defined region of the genome, DNA fragments or subclones for germ-line transformation or targeted gene interruption (Ballinger and Benzer 1989), and especially substrates for genomic or cDNA sequencing.

ACCESS TO MATERIALS AND RELATIONS WITH OTHER *DROSOPHILA* GENOME PROJECTS

The clones and information generated in the course of this mapping project are available to *Drosophila* researchers on request of the author and may be used for any research purpose without restriction. As of January, 1992, we have provided more than 300 clones of various regions to more than 100 investigators whose gene of interest is covered by a YAC.

Our approach to the *Drosophila* genome is part of a larger international effort that includes a European consortium of M. Ashburner, D. M. Glover, F. Kafatos, C. Louis, R.D.C. Saunders, I. Sidén-Kiamos, and other collaborators. These investigators have adopted a cosmid-based approach that uses probes obtained from microdissected salivary gland chromosomes to identify cosmids in the region, which are then assembled into contigs by means of fingerprinting (Sidén-Kiamos et al. 1990; Kafatos et al. 1991). A graphic summary of the results of this project, as of July 1,1991, can be found in the fold-out poster in the October 11, 1991, issue of *Science* (254: 165–344).

Another *Drosophila* group is headed by H. Lehrach and J. Hoheisel, who have relied on hybridization with synthetic oligonucleotides to identify clones having the oligonucleotide sequences in common in a number of libraries, including our YAC and P1 libraries, the cosmid libraries of the European consortium, lambda libraries, cDNA libraries, and so on (Lehrach et al. 1990; Hoheisel and Lehrach 1992).

The various strategies for *Drosophila* genome mapping are complementary and share the goal of producing a high-resolution physical map of the genome that can be used to isolate, sequence, and identify all of the genes in the organism. Achieving this goal would be the climax of the phase of *Drosophila* genetics opened up 80 years ago by T.H. Morgan. More importantly, it would represent the beginning of a new synthesis of biochemistry, physiology, cell biology, development, and behavior, in which all the relevant genes and their products will be known and the emphasis will be on how the organism functions as an integrated whole.

Acknowledgments

Many colleagues and collaborators have contributed to making this work possible. Particularly important has been the contribution of Ian W. Duncan and his collaborators, H. Cai, P. Kiefel, and J. Yee, who carried out many of the YAC in situ hybridizations and generously provided the data for inclusion in the present analysis. Contributors from my own laboratory include J.W. Ajioka, D. Garza, R.W. Jones, M.A. Lackey, A.R. Lohe, E.R. Lozovskaya, H. Ochman, D. Petrov, D.A. Smoller, and A.E.C. Vellek. Other important contributions have been made by O.N. Danilevskaya, A.J. Link, C.H. Martin, C. Mayeda, and M.J. Palazzolo. Allan R. Lohe made important suggestions for improving the manuscript. This work was supported by grant HG-00357 from the National Center for Human Genome Research.

References

Adams, M.D., J.M. Kelley, J.D. Gocayne, M. Dubnick, M.H. Polymeropoulos, H. Xiao, C.R. Merril, A. Wu, B. Olde, R.F. Moreno, A.R. Kerlavage, W.R. McCombie, and J.C. Venter. 1991. Complementary DNA sequencing: Expressed sequence tags and human genome project. *Science* **252:** 1651.

Ajioka, J.W., D. Garza, D. Johnson, J.P. Carulli, R.W. Jones, and D.L. Hartl. 1990. Genome evolution analyzed by cloning large fragments of *Drosophila* DNA in yeast artificial chromosomes. In *Molecular evolution* (ed. M.T. Clegg and S.J. O'Brien), p. 253. Alan R. Liss, New York.

Ajioka, J.W., D.A. Smoller, R.W. Jones, J.P. Carulli, A.E.C. Vellek, D. Garza, A.J. Link, I.W. Duncan, and D.L. Hartl. 1991. *Drosophila* genome project: One-hit coverage in yeast artificial chromosomes. *Chromosoma* **100:** 495.

Ashburner, M. 1989. Drosophila: *A laboratory handbook*. Cold Spring Harbor Laboratory Press, Cold Spring Harbor, New York.

Ballinger, D.G. and S. Benzer. 1989. Targeted gene mutations in *Drosophila*. *Proc. Natl. Acad. Sci.* **86:** 9402.

Barillot, E., J. Dausset, and D. Cohen. 1991. Theoretical analysis of a physical mapping strategy using random single-copy landmarks. *Proc. Natl. Acad. Sci.* **88:** 3917.

Beermann, W. 1972. Chromomeres and genes. In *Developmental studies on giant chromosomes* (ed. W. Beermann et al.), p. 1. Springer-Verlag, New York.

Bingham, P.M., R. Levis, and G.M. Rubin. 1981. Cloning of DNA sequences from the white locus of *Drosophila melanogaster* by a novel and general method. *Cell* **25:** 693.

Bridges, C.B. 1935. Salivary chromosome maps with a key to the banding of the chromosomes of *Drosophila melanogaster*. *J. Hered.* **26:** 60.

Burke, D.T., G.F. Carle, and M.V. Olson. 1987. Cloning of large segments of exogenous DNA into yeast by means of artificial chromosome vectors. *Science* **236:** 806.

Carle, G.F. and M.V. Olson. 1984. Separation of chromosomal DNA molecules from yeast by orthogonal-field-alternation gel electrophoresis. *Nucleic Acids Res.* **12:** 5647.

Carlson, E.A. 1966. *The gene: A critical history*. W.B. Saunders, Philadelphia.

Cooley, L., C. Berg, and A. Spradling. 1988a. Controlling *P* element insertional mutagenesis. *Trends Genet.* **4:** 254.

Cooley, L., R. Kelley, and A. Spradling. 1988b. Insertional mutagenesis of the *Drosophila* genome with single P elements. *Science* **239:** 1121.

Coulson, A., J. Sulston, S. Brenner, and J. Karn. 1986. Toward a physical map of the genome of the nematode *Caenorhabditis elegans*. *Proc. Natl. Acad. Sci.* **83:** 7821.

Coulson, A., R. Waterston, J. Kiff, J. Sulston, and Y. Kohara. 1988. Genome linking with yeast artificial chromosomes. *Nature* **335:** 184.

Coulson, A., Y. Kozono, B. Lutterbach, R. Shownkeen, J. Sulston, and R. Waterston. 1991. YACs and the *C. elegans* genome. *BioEssays* **13:** 413.

Danilevskaya, O.N., E.V. Kurenova, M.N. Pavlova, A.J. Link, A. Koga, A. Vellek, and D.L. Hartl. 1990. He-T family DNA sequences in the Y chromosome of *Drosophila melanogaster* share homology with the X-linked *Stellate* genes. *Chromosoma* **100:** 118.

Engels, W.R. 1989. P elements in *Drosophila melanogaster*. In *Mobile DNA* (ed. D. E. Berg and M. Howe), p. 437. American Society for Microbiology, Washington, D.C.

Feinberg, A.P. and B. Vogelstein. 1984. A technique for radiolabeling DNA restriction endonuclease fragments to high specific activity: Addendum. *Anal. Biochem.* **137:** 266.

Garza, D., J.W. Ajioka, D.T. Burke, and D.L. Hartl. 1989a. Mapping the *Drosophila* genome with yeast artificial chromosomes. *Science* **246:** 641.

Garza, D., J.W. Ajioka, J.P. Carulli, R.W. Jones, D.H. Johnson, and D.L. Hartl. 1989b. Physical mapping of complex genomes. *Nature* **340:** 577.

Green, E.D. and M.V. Olson. 1990a. Chromosomal region of the cystic fibrosis gene in yeast artificial chromosomes: A model for human genome mapping. *Science* **250:** 94.

————. 1990b. Systematic screening of yeast artificial-chromosome libraries by use of the polymerase chain reaction. *Proc. Natl. Acad. Sci.* **87:** 1213.

Hartl, D.L. and M.J. Palazzolo. 1992. *Drosophila* as a model organism in genome analysis. In *Techniques and applications of genome research* (ed. K.W. Adolph). Academic Press, Orlando, Florida. (In press.)

Hartl, D.L., J.W.Ajioka, H. Cai, A.R. Lohe, E.R. Lozovskaya, D.A. Smoller, and I.W. Duncan. 1992. Towards a *Drosophila* genome map. *Trends Genet.* **8:** 70.

Heino, T.I., A.O. Saura, and V. Sorsa. 1992. Maps of the salivary gland chromosomes of *Drosophila melanogaster*. *Drosophila Inf. Serv.* (in press).

Hieter, P., C. Connelly, J. Shero, M.K. McCormick, S. Antonarakis, W. Pavav, and R. Reeves. 1990. Yeast artificial chromosomes: Promises kept and pending. In *Genome analysis volume 1: Genetic and physical mapping* (ed. K.E. Davies and S.M. Tilghman), p. 83. Cold Spring Harbor Laboratory Press, Cold Spring Harbor, New York.

Hoheisel, J. and H. Lehrach. 1992. Relational genome analysis of *Drosophila melanogaster*. *Trends Genet.* (in press).

Kafatos, F.C., C. Louis, C. Savakis, D.M. Glover, M. Ashburner, A.J. Link, I. Sidén-Kiamos, and R.D.C. Saunders. 1991. Integrated maps of the *Drosophila* genome: Progress and prospects. *Trends Genet.* **7:** 155.

Langer-Safer, P.R., M. Levine, and D.C. Ward. 1982. Immunological method for mapping genes on *Drosophila* polytene chromosomes. *Proc. Natl. Acad. Sci.* **79:** 4381.

Lefevre, G., Jr. 1976. A photographic representation and interpretation of the polytene chromosomes of *Drosophila melanogaster* salivary glands. In *The genetics and biology of* Drosophila (ed. M. Ashburner and E. Novitski), p. 31. Academic Press, New York.

Lehrach, H., R. Drmanac, J. Hoheisel, Z. Larin, C. Lennon, A.P. Monaco, D. Nizetic, G. Zehetner, and A. Poustka. 1990. Hybridization fingerprinting in genome mapping and sequencing. In *Genome analysis volume 1: Genetic and physical mapping* (ed. K.E. Davies and S.M. Tilghman), p. 39. Cold Spring Harbor Laboratory Press, Cold Spring Harbor, New York.

Livak, K.J. 1990. Detailed structure of the *Drosophila melanogaster Stellate* genes and their transcripts. *Genetics* **124:** 303.

Lohe, A.R. and D.L. Brutlag. 1986. Multiplicity of satellite DNA sequences in *Drosophila melanogaster*. *Proc. Natl. Acad. Sci.* **83:** 696.

Merriam, J., M. Ashburner, D.L. Hartl, and F.C. Kafatos. 1991. Toward cloning and mapping the genome of *Drosophila*. *Science* **254:** 221.

Miklos, G.L.G., M. Yamamoto, J. Davies, and V. Pirrotta. 1988. Microcloning reveals a high frequency of repetitive sequences characteristic of chromosome 4 and the β-heterochromatin of *Drosophila melanogaster*. *Proc. Natl. Acad. Sci.* **85:** 2051.

Mlodzik, M., Y. Hiromi, U. Weber, C.S. Goodman, and G.M. Rubin. 1990. The *Drosophila* seven-up gene, a member of the steroid receptor gene super-family, controls photoreceptor cell fates. *Cell* **60:** 211.

Ochman, H., M.M. Medhora, D. Garza, and D.L. Hartl. 1990. Amplification of flanking sequences by inverse PCR. In *PCR protocols: A guide to methods and applications* (ed. M. Innis et al.), p. 219. Academic Press, New York.

Olson, M.V., L. Hood, C. Cantor, and D. Botstein. 1989. A common language for physical mapping of the human genome. *Science* **245:** 1434.

Painter, T.S. 1933. A new method for the study of chromosome rearrangements and the plotting of chromosome maps. *Science* **78**: 585.

Palazzolo, M.J., D.R. Hyde, K. VijayRaghavan, K. Mecklenberg, S. Benzer, and E. Meyerowitz. 1989. Use of a new strategy to isolate and characterize 436 *Drosophila* cDNA clones corresponding to RNAs detected in adult heads but not early embryos. *Neuron* **3**: 527.

Palazzolo, M.J., S. Sawyer, C.H. Martin, D.A. Smoller, and D.L. Hartl. 1991. Optimized strategies for sequence-tagged-site selection in genome mapping. *Proc. Natl. Acad. Sci.* **88**: 8034.

Pierce, J.C. and N.L. Sternberg. 1992. Using the bacteriophage P1 system to clone high molecular weight (HMW) genomic DNA. *Methods Enzymol.* (in press).

Pierce, J.C., B. Sauer, and N. Sternberg. 1992. A positive selection vector for cloning of high molecular weight DNA by the bacteriophage P1 system: Improved cloning efficacy. *Proc. Natl. Acad. Sci.* **89**: 2056.

Roberts, D.B. 1986. Drosophila: *A practical approach.* IRL Press, Washington, D.C.

Rudkin, G.T. 1961. Cytochemistry in the ultraviolet. *Microchem. J. Symp. Ser.* **1**: 261.

———. 1969. Non replicating DNA in *Drosophila. Genetics* (suppl.) **61**: 227.

Sidén-Kiamos, I., R.D.C. Saunders, L. Spanos, T. Majerus, J. Treanear, C. Savakis, C. Louis, D.M. Glover, M. Ashburner, and F.C. Kafatos. 1990. Towards a physical map of the *Drosophila melanogaster* genome: Mapping of cosmid clones within defined genomic divisions. *Nucleic Acids Res.* **18**: 6261.

Smoller, D.A., D. Petrov, and D.L. Hartl. 1991. Characterization of bacteriophage P1 library containing inserts of *Drosophila* DNA of 75-100 kilobase pairs. *Chromosoma* **100**: 487.

Sorsa, V. 1988. *Chromosome maps of* Drosophila. CRC Press, Boca Raton, Florida.

Steller, H. and V. Pirrotta. 1985. A transposable *P* vector that confers selectable G418 resistance to *Drosophila* larvae. *EMBO J.* **4**: 167.

Sternberg, N. 1990. Bacteriophage P1 cloning system for the isolation, amplification, and recovery of DNA fragments as large as 100 kilobase pairs. *Proc. Natl. Acad. Sci.* **87**: 103.

Sturtevant, A.H. 1913. The linear arrangement of six sex-linked genes in *Drosophila*, as shown by their mode of association. *J. Exp. Zool.* **14**: 43.

———. 1965. *A history of genetics.* Harper & Row, New York.

Traverse, K.L. and M.L. Pardue. 1989. Studies of He-T DNA sequences in the pericentric regions of *Drosophila* chromosomes. *Chromosoma* **97**: 261.

Vogelstein, B. and D. Gillespie. 1979. Preparative and analytical purification of DNA from agarose. *Proc. Natl. Acad. Sci.* **76**: 615.

Watson, J.J. and R.M. Cook-Deegan. 1991. Origins of the human genome project. *FASEB J.* **5**: 8.

Wu, C.-I., T.W. Lyttle, M.-L. Wu, and G.-F. Lin. 1988. Association between a satellite DNA sequence and the *Responder of Segregation Distorter* in D. *melanogaster. Cell* **54**: 179.

Physical Mapping of the *Arabidopsis thaliana* Genome

Renate Schmidt and Caroline Dean

Department of Molecular Genetics, Cambridge Laboratory, John Innes Centre
AFRC Institute of Plant Science Research
Norwich NR4 7UJ, United Kingdom

Arabidopsis thaliana (Thale cress, Arabidopsis) has been adopted by a large number of plant biologists as the model organism in which to dissect many plant processes using a molecular genetic approach. The technologies required for efficient map-based cloning strategies have advanced rapidly and, in the next few years, the number of genes cloned from Arabidopsis using chromosome walking, insertional mutagenesis, or subtractive hybridization techniques will increase enormously. This concentration of effort on one plant species with a small genome size makes the construction of a complete physical map of the genome an enticing and achievable goal for the Arabidopsis community. The advantages of having a physical map are many and outweigh the laborious process required to generate it. The physical map will greatly facilitate gene isolation without the need for each laboratory to initiate chromosome walks. The map will also provide the basis for understanding the organization of a plant genome. New genes can be identified, and analysis can be carried out of the physical linkage of the genes, the distribution of repetitive elements, the identification of centromeric and sub-telomeric sequences, and the distribution of recombination points that dictate the relationship of genetic and physical distance.

In this chapter we describe:

❏ the generation of local physical maps from chromosome walking experiments

❏ progress in the construction of an overlapping library of the Arabidopsis genome in cosmid and yeast artificial chromosome (YAC) clones

Genome Analysis Volume 4: *Strategies for Physical Mapping*
© 1992 Cold Spring Harbor Laboratory Press 0-87969-412-2/92 $3 + 00

❑ specific examples of our work linking YAC clones on chromosomes 4 and 5 and the identification of YAC clones carrying repeated DNA sequences

❑ future strategies for completing the physical map

INTRODUCTION

Many processes fundamental to the development and functioning of flowering plants remain very poorly understood. Of special interest are processes such as the transition from vegetative development to flowering, the perception of and reaction to environmental stimuli, and the molecular recognition involved in plant-pathogen interactions. Until recently, these processes had not been approached using a molecular genetic analysis. This situation, however, is changing rapidly. There is now considerable emphasis on the isolation of genes using molecular genetic approaches in species such as maize, snapdragon, tomato, wheat, rice, and most notably, Arabidopsis. Arabidopsis has many advantages for such an analysis:

1. It is a very small plant with a rapid generation time (6 weeks from seed to seed). This makes large mutagenesis screens quite feasible. This plant also produces large numbers (up to 10,000) of seed.
2. The haploid genome size is estimated to be approximately 100 Mb (Hwang et al. 1991), similar to that of the nematode *Caenorhabditis elegans*. This is considerably smaller than most other well-studied plant systems (see Table 1). It also has a very low level of repeated DNA sequences (Leutwiler et al. 1984; Pruitt and Meyerowitz 1988).

Table 1 Haploid genome sizes for various plant species

Plant species	Mb/1C
Arabidopsis thaliana (Arabidopsis)	70,[a] 100,[b] 145[c]
Oryza sativa (rice)	415–463[c]
Lycopersicon esculentum (tomato)	907–1,000[c]
Solanum tuberosum (potato)	1,597–1,862[c]
Zea mays (maize)	2,292–2,716[c]
Nicotiana tabacum (tobacco)	4,221–4,646[c]
Hordeum vulgare (barley)	4,873[c]
Triticum aestivum (wheat)	15,966[c]

[a]Leutwiler et al. (1984).
[b]Hwang et al. (1991).
[c]Arumuganathan and Earle (1991).

3. There are two published restriction fragment length polymorphism (RFLP) maps (Chang et al. 1988; Nam et al. 1989) which between them have a total of nearly 200 mapped markers. Since publication, about another 160 markers have been added. This translates to an average distance from a known marker to any point in the genome of around 70 kb (Meyerowitz et al. 1991). Randomly amplified polymorphic DNA (RAPD) markers (Williams et al. 1990) are also being employed, and currently 252 of these markers are mapped onto the Arabidopsis genetic map (Reiter et al. 1992).

4. A transformation system using the soil bacterium *Agrobacterium tumefaciens* is available. Many *Agrobacterium* vector systems using a number of different antibiotic or herbicide resistance markers have been developed to select for transformants carrying the integrated T-DNA (the part of the *A. tumefaciens* Ti plasmid that is integrated into the plant chromosome). A transformation procedure where a culture of *A. tumefaciens* is mixed with root explants (Valvekens et al. 1988), followed by regeneration of intact plants from the transformed root material, is being used by many laboratories. A non-tissue-culture procedure based on imbibing seed in a solution of *Agrobacterium* has also been successful for Feldmann and Marks (1987).

A large number of Arabidopsis mutants have been generated using a variety of mutagens. These have been used by Koornneef et al. (1983) to establish a genetic map which now has 117 markers distributed over the five chromosomes. The map positions of the loci involved in floral induction or response to the plant hormone, gibberellic acid, are shown in Figure 1. The number of newly identified genetic loci in Arabidopsis is increasing at a considerable rate. The classes of mutants include (1) physiological, e.g., hormonal, flowering time, photoreception, pollen-stigma interactions, mineral-ion absorption, pathogen interactions, anthocyanin production; (2) metabolic, e.g., lipid and starch biosynthesis, nitrate metabolism and auxotrophs; and (3) morphological, e.g., flower structure, leaf shape, trichome development, root development, embryo development.

A large new collection of embryo- (>25,000) and seedling- (>5,000) lethal mutations has recently been described (Mayer et al. 1991). This resulted from a systematic search for mutations that disrupted the spatial organization of the seedling by altering embryogenesis. Multiple alleles of nine loci involved in Arabidopsis body organization were identified. The size of the screen was so large that mutations were potentially identified in at least one-sixth of the Arabidopsis genome (Meyerowitz et al. 1991).

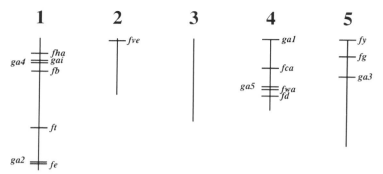

Figure 1 Distribution of late flowering and dwarfing genes in *A. thaliana*. The figure shows the approximate map position according to Koornneef (1990) of the late flowering (*f..*) and dwarfing (gibberellic acid deficient, *ga1-ga5*/gibberellic acid insensitive, *gai*) loci on the five Arabidopsis chromosomes.

Newly identified loci are being mapped, either relative to other visible markers, using the 19 mapping strains with multiple markers on each chromosome developed by Koornneef (Koornneef et al. 1987; Meyerowitz et al. 1991), or relative to RFLP or RAPD markers.

GENERATION OF THE ARABIDOPSIS PHYSICAL MAP

Local physical maps from chromosome walking experiments

The increasing density of RFLP/RAPD markers and the small genome size with relatively few repeated DNA sequences mean that gene isolation using chromosome walking is a feasible prospect in Arabidopsis, unlike most other plant species. During the last couple of years, many chromosome walking experiments have been initiated, with the aim of cloning various mutant loci. These walks are generating local physical maps at various regions around the genome. The walks initially used cosmid libraries (Bleecker 1991; Hauge et al. 1991) constructed in the cosmid vectors, pCIT30 (Yanofsky et al. 1990) or Lorist (Cross and Little 1986). More recently, the availability of several YAC libraries has meant that most chromosome walking experiments employ YAC clones, since the inserts are on average three times larger than the inserts in the cosmid clones. More than 30 chromosome walking projects in over 20 laboratories are currently known to be in progress (Meyerowitz et al. 1991). Some examples are summarized in Table 2.

The general strategy that has been adopted for chromosome walking experiments in Arabidopsis is to generate a YAC contig covering the region containing the locus of interest, in combination with fine mapping of the locus within that region. One strategy for mapping the locus is outlined in Figure 2. The mutant locus to be cloned (**b**) is initially

Table 2 Examples of chromosome walking experiments toward specific mutant loci

Locus	Chromosome	Reference
abi1 (abscisic acid insensitivity)	4	Grill and Somerville (1991b); Leung et al. (1991
abi2 (abscisic acid insensitivity)		Grill and Somerville (1991b)
abi3 (abscisic acid insensitivity)	3	Leung et al. (1991)
axr1 (auxin resistance)	1	Leyser et al. (1991)
axr2 (auxin resistance)	3	Wilson et al. (1990)
etr (ethylene resistance)	1	Bleecker (1991)
ga2 (gibberellic acid deficiency)	1	Hauge et al. (1991)
gai (gibberellic acid insensitivity)	1	Hauge et al. (1991)
det1 (de-etiolated)	4	Delaney et al. (1991)
det2 (de-etiolated)	2	Nagpal and Chory (1991)
fca (late flowering)	4	Westphal et al. (1991)
fg (late flowering)	5	Putterill et al. (1991)
fve (late flowering)	2	Cruz-Alvarez et al. (1991)
fwa (late flowering)	4	Koornneef et al. (1991)
gi2 (late flowering)	1	Araki et al. (1989)
fae1 (deficiency in fatty acid elongation)	4	Lemieux et al. (1991)
ttg (glabrous, lack of anthocyanins)	5	Walker and Gray (1991)
ara1 (arabinose sensitivity)	4	Medd et al. (1991)
ms1 (male sterility)	5	Chaudhury et al. (1991); Wilson et al. (1991)
rpm1 (resistance to a *Pseudomonas* pathovar)	3	Debener et al. (1991)

mapped relative to other visible markers (e.g., A and C). These flanking markers are then used to select recombination events close to **b** in F_2 progeny from a cross between two parents (e.g., Landsberg *erecta* and Columbia) which are polymorphic at the DNA level. DNA is isolated from pooled progeny plants carrying recombination events between **b** and A or **b** and C. RFLP markers mapping to the same region are hybridized to Southern blots of DNA from these recombinants. Recombination points are mapped relative to these DNA markers by monitoring the switch of the fragment pattern on the Southern blots from a homozygous pattern to a heterozygous pattern (i.e., both Landsberg *erecta* and Columbia polymorphisms present). The analysis of the recombination breakpoints determines the map position of **b** with respect to the RFLP markers.

YAC clones corresponding to the RFLP markers flanking **b** are then identified by using the RFLP markers as probes on YAC libraries (discussed below). End probes from these YAC clones are generated and hybridized back to the YAC libraries to find overlapping YAC clones. The direction of the walk toward **b** is determined by utilizing the recombination events between A and **b** or C and **b** and end probes from the YAC clones that reveal polymorphisms between Landsberg *erecta* and Columbia. Recombination points again are mapped relative to these DNA markers by monitoring the switch from the homozygous to the heterozygous pattern. Once the walks are oriented, they are extended toward **b** in order to build one contiguous region covering **b**.

As soon as the locus of interest has been localized to one or part of one YAC clone, DNA from this YAC clone is subcloned into vectors,

Figure 2 Fine-mapping of a locus. This figure demonstrates one strategy to fine-map a locus of interest. (*a*) Crosses are made to isolate recombinant individuals between the locus of interest **b** and the flanking loci A and C. These crosses can be set up with either the flanking markers in coupling (as shown in the figure) or in repulsion (aa**BB** x AA**bb**). The mutations in coupling in the Landsberg *erecta* background (black lines) are crossed to a wild-type line of a different ecotype, namely Columbia (stippled line), which is polymorphic at the DNA level. Resulting F_1 plants are allowed to self-pollinate, and recombinant individuals are selected for in the F_2 generation. In the figure only one class of possible recombinants is indicated. The recombinant individuals are subjected to an RFLP analysis, the result of which is shown in *b*. Recombination breakpoints for five different recombinants are depicted, although it should be noted that the breakpoints can only be defined to an area between two markers in such an analysis. The distribution of the recombination events in the different individuals allows the RFLP markers 1, 2, and 3 used in this analysis to be mapped with respect to the loci A, **b**, and C. YAC clones corresponding to the RFLP markers 2 and 3 are isolated (*c*). End fragments of some of these YAC clones are generated to serve as probes for the isolation of overlapping YAC clones to form a contiguous region covering the locus. Furthermore, end probes α, β, γ that reveal an RFLP between the two Arabidopsis ecotypes are used to obtain a higher resolution map position of **b** relative to the YAC end fragments (*d*) using recombinant individuals. This type of analysis defines the position of the locus of interest to a small genetic region.

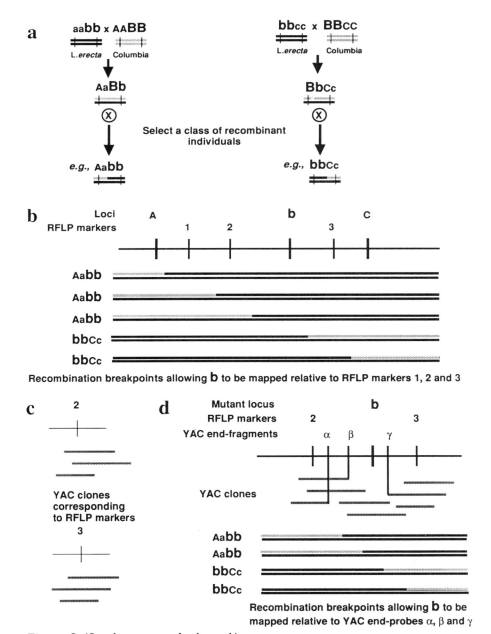

Figure 2 (*See facing page for legend.*)

which can be used in plant transformation experiments. This can be achieved by subcloning the YAC directly or using it as a probe to a library of Arabidopsis DNA cloned in a plant transformation vector. The subclones are then used to transform the mutant Arabidopsis line and identify clones that complement the mutant phenotype. Final identification of the open reading frame is through further rounds of complementation with smaller subclones, in conjunction with sequence and transcript analysis. Some of the ongoing chromosome walks have nearly reached a conclusion. Several have localized the locus to one cosmid clone and are in the process of final rounds of complementation with transcript and sequence analysis.

Progress in the construction of an overlapping cosmid library of the Arabidopsis genome

To facilitate future cloning in Arabidopsis, Hauge et al. (1991) initiated a project several years ago to generate an overlapping cosmid library of the Arabidopsis genome. They adopted the same nondirected strategy used for the *C. elegans* genome project by Coulson, Sulston, and co-workers (Coulson et al. 1986). Random cosmid clones (average size 40 kb) were "fingerprinted" by digestion with *Hind*III (which cut each cosmid an average of 15 times), labeled with [^{32}P]dATP, and subsequently cleaved with *Sau*3A. The labeled fragments were then separated by polyacrylamide gel electrophoresis, and the banding patterns were entered into the computer using a scanning densitometer and an image-processing package (Sulston et al. 1989). The regions of overlap were determined, and the clones were assembled into contigs (Coulson et al. 1986; Sulston et al. 1988). After analysis of 17,000 clones, 750 contigs have been assembled (see Table 3). Cosmid contigs cover 91–95 % of the genome. On the basis of this work, Hauge and Goodman estimate the genome size to be approximately 100,000 kb (Hwang et al. 1991). Leutwiler et al. (1984) initially gave an estimate of the Arabidopsis genome size of 70,000 kb, based on studies of reassociation kinetics. Recently, an even higher value (145,000 kb) was proposed after studies using flow cytometry (Arumuganathan and Earle 1991).

Taking into account a genome size of 100,000 kb and a 91–95% coverage of the genome in cosmid contigs, the average size of the identified cosmid contigs is approximately 120–130 kb. The number of cos-

Table 3 Current status of the overlapping cosmid library

Number of clones "fingerprinted" and analyzed	17,000
Sampling redundancy	8- to 10-fold
Number of contigs obtained	750
Percentage of genome represented in contigs	91–95

Data from Hauge et al. (1991).

mids fingerprinted represent 8–10 genome equivalents, so a continuation of this approach to reduce the number of contigs is no longer fruitful. The difficulties of obtaining map closure utilizing cosmid clones are thought to be caused by regions of eukaryotic genomes being unclonable in *Escherichia coli* (Coulson et al. 1991). It may also be due to the relatively large overlap (35%) required before a cosmid is included in a contig or the presence of clones that contain few or no *Hin*dIII sites, since a minimal number of bands is required in the statistical analysis to detect overlapping clones (Hauge et al. 1991). Thus, reducing the number of cosmid contigs depends on adopting different approaches. Hauge et al. (1991) have again followed the *C. elegans* model and are hybridizing YAC clones to representative arrays of the cosmid contigs. Utilization of YAC clones to link the cosmid contigs has proved very effective at reducing the number of contigs covering the *C. elegans* genome (Coulson et al. 1988; Coulson et al. 1991). The YAC clones provide two advantages over cosmids. First, YAC clones carry much larger inserts, and second, it is thought that there will be fewer unclonable sequences in yeast than in *E. coli* (Coulson et al. 1991).

Generation of an overlapping YAC library of the Arabidopsis genome

In addition to the effort described above, there is a large international collaboration to generate an overlapping YAC library of the Arabidopsis genome in a more directed manner. This project has taken advantage of the relatively extensive framework of mapped markers and repeated DNA sequences to position YAC clones on the genetic map (Meyerowitz et al. 1991). The YAC clones are then being linked in directed walking experiments. Generation of YAC contigs in addition to cosmid contigs will serve several purposes. First, it will help to ensure an accurate and eventually a complete physical map. It will also produce a selection of different-sized clones throughout the genome that can be used for a variety of different experiments. In the future, mapped P1 clones, generated via a cloning system based on the bacteriophage P1, may also contribute to this resource (J. Ecker, pers. comm.). The YAC contigs may also carry the sequences not found in cosmid contigs. Last, a representative set of YAC clones, arrayed in order, could provide a minimal set of clones covering the Arabidopsis genome with which to accurately fine-map new probes.

It will be important to effectively coordinate all the physical mapping efforts. One plan to achieve this is to hybridize end probes from the YAC walks onto the filters carrying representative arrays of the cosmid contigs.

We (C. Dean and co-workers, Norwich) are focusing on linking YAC clones corresponding to the top halves of chromosomes 4 and 5. The Goodman laboratory (Massachusetts General Hospital, Boston) is focusing on chromosomes 2 and 3, and the Ecker (University of Philadelphia) and Scolnik (DuPont) laboratories are focusing on chromosome 1. Many other groups are contributing to this effort through their chromosome walking experiments to different loci distributed over the genome (summarized in Table 2).

The three YAC libraries used in the initial stages of this project were constructed from DNA isolated from the Arabidopsis ecotype Columbia or the Landsberg *erecta* mutant *abi-1* and fractionated using either *Bam*HI partial digestion (EG and *abi* libraries, Grill and Somerville 1991a) or random shear of high-molecular-weight Arabidopsis DNA (EW library, Ward and Jen 1990). A derivative of pYAC4 (Burke et al. 1987), pYAC41, carrying T3 bacteriophage promoters flanking a *Bam*HI cloning site, was the vector used for two of the libraries (Grill and Somerville 1991a). Each of these YAC libraries contains between 2100 and 2300 YAC clones. The average insert size has been established to be approximately 150 kb for the two Columbia libraries. Therefore, these libraries should contain approximately three genome equivalents. A new library of 2300 clones, constructed using *Eco*RI partial digests cloned into pYAC4 (Burke et al. 1987), has recently become available, and this has an average insert size of 250 kb (Ecker 1990).

The strategy adopted by the different research groups involved in the linking of the YAC clones was to hybridize all the available RFLP markers (Chang et al. 1988; Nam et al. 1989) in the region being focused on by each group to at least one of the available YAC libraries. After YAC clones corresponding to all mapped markers were identified, walks would then be initiated to link up the YAC clones hybridizing to adjacent RFLP markers. Since more than 360 RFLP markers are available, with, on average, one every 1–2 cM in the genome, the average distance between each of the YAC clones hybridizing to adjacent RFLP markers would be approximately 150–300 kb or 1–2 steps.

Table 4 summarizes results from participating laboratories, which were compiled and presented by Hwang et al. (1991) in April 1991, 18

Table 4 Current status of the overlapping YAC library

Number of RFLP markers used as probes	125
Number of YAC clones identified	296
Number of RFLP markers shown to be physically linked by one or more YACs	32
Number of YAC contigs spanning two or more RFLP markers	12
Average insert size of analyzed YAC clones (196)	160 kb
Average size of YAC contigs	240 kb
Percentage of genome covered by mapped YACs	30

Data from Hwang et al. (1991).

months after the international collaboration was initiated. This effort resulted in approximately 30% of the Arabidopsis genome being represented in mapped YAC clones. Furthermore, a number of RFLP markers (25% of the markers tested) were shown to be physically linked by one or more YAC clones. The physically linked RFLP markers (32) could so far be placed in 12 contigs encompassing 2–5 RFLP markers. Since then, many more markers have been hybridized to the libraries.

SPECIFIC EXAMPLES FROM LINKING YAC CLONES ON CHROMOSOMES 4 AND 5

Efficient screening of the YAC libraries

Our first priority upon starting the project to link the YAC clones covering the top halves of chromosomes 4 and 5 was to maximize the efficiency with which we could screen the libraries. Colonies were screened using the yeast colony hybridization technique of Coulson et al. (1988). Utilizing a 96-prong replicator and offsetting each colony print slightly, we can now plate 24 x 96 colonies (= 2304, i.e., one whole library) on one plate which fits onto an 11-cm x 8-cm nylon filter. Each of these master filters can serve to produce multiple slave-copies (20–50) as described by Coulson et al. (1988). A schematic illustration of the 24 offsets and a typical hybridization result using such a membrane carrying one of the Columbia libraries is shown in Figure 3. The hybridization and washing conditions were established so that a low level of background hybridization remained on the filters to identify all the colonies. This ensured that the coordinates of the positively hybridizing YAC clones could easily be determined. To confirm the identification of the positively hybridizing colonies, duplicate filters carrying different offset arrays are always used.

In our laboratory, all 43 available RFLP markers, mapping to the top halves of chromosomes 4 and 5, have now been hybridized to the YAC libraries. Examples of positively hybridizing YAC colonies to two of these RFLP markers are shown in Figure 4. The RFLP marker 6833 mapped on one RFLP map (Nam et al. 1989) and the chalcone synthase probe *chs2* (Feinbaum and Ausubel 1988) mapped on the other RFLP map (Chang et al. 1988) hybridize to the YAC clones EG12F8 and EG18H3 (Fig. 4). Therefore, the YAC clones EG12F8 and EG18H3 span both markers. Markers 6833 and *chs2* thus provide a contact point between the two RFLP maps. Furthermore, the CHS2 gene has been shown to correspond to the anthocyanin-deficient mutant *tt4* (G. Coupland, unpubl.), and so a link between both RFLP maps and the visible marker map can be established at this locus (Fig. 5).

All the positively hybridizing YAC colonies are subjected to two further analyses. First, restriction-digested yeast DNA is analyzed along-

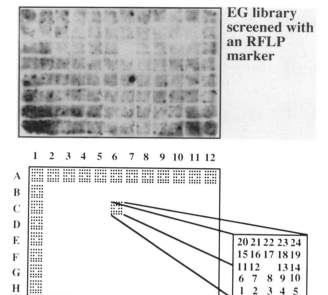

Figure 3 High-density yeast colony screening. Example of a yeast colony hybridization using an RFLP marker as probe. The EG library (Grill and Somerville 1991a) has been arrayed using 24 offsets of the 96-prong applicator; the filter thus represents the complete library. The schematic representation shows the details of the high-density plating.

side the RFLP marker DNA by Southern blot analysis using the RFLP marker as a probe. This verifies the colony hybridization results and, additionally, provides information about potential rearrangements within the YAC insert and whether the YAC clones span the RFLP marker completely. Second, it is necessary to demonstrate that the YAC clone contains the fragment that was scored as polymorphic in the RFLP mapping experiments. This is to establish that the YAC clone maps to the same position as the RFLP marker. If the mapped polymorphism has a fragment size identical to the one observed in the RFLP marker DNA and the YAC clone shares all fragments with the RFLP marker, unequivocal linkage is then established. However, in cases where the RFLP marker does not show a fragment of the same size as the polymorphic band that was mapped, it is necessary to determine that the corresponding YAC clone contains the polymorphism.

The groups involved in the international collaboration have predominantly used the EG YAC library of the Columbia ecotype (Grill and Somerville 1991a; Hwang et al. 1991). However, we have now completed the comparative analysis of the EG and EW (Ward and Jen 1990) libraries using all the RFLP markers mapping to the top halves of chromosomes 4 and 5. Many of them (13 out of 23 markers for chromosome 5) identify YAC clones in both libraries (EG and EW), but

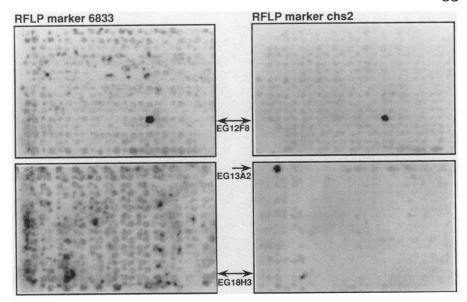

Figure 4 YAC clones link RFLP markers 6833 and *chs2*. The colony hybridization of identical filters of the EG library (Grill and Sole 1991a) (in this case containing 384 YAC clones –4 offsets) is shown for the RFLP markers 6833 (Nam et al. 1989) and *chs2* (Chang et al. 1988). Arrows indicate the positively hybridizing YAC clones.

5 markers gave positives only in the EG library and 4 only in the EW library (cf. Fig. 5). For RFLP marker 224, we were not able to identify YAC clones in either of the two Columbia libraries, but we could identify two YAC clones in the Landsberg *erecta abi* library (Hwang et al. 1991). We have generally only hybridized markers to the *abi* library when positives were not found in either of the two Columbia libraries. The differential representation of various regions of the genome in the three libraries, all of which should contain (in theory) three genome equivalents, shows how important it is to work with multiple YAC libraries. Some of the nonrepresentation of YAC clones in the EG library may reflect the unequal distribution of *Bam*HI sites in the Arabidopsis genome, as a *Bam*HI partial digest was used in the construction of the library. However, up until now, YAC clones corresponding to all RFLP markers tested have been identified in at least one of the libraries (Hwang et al. 1991).

Figure 5 summarizes the YAC coverage on the top half of chromosome 5 along with the visible marker map. The boxes correspond to a YAC contig around each RFLP marker on chromosome 5. Because the two published RFLP maps are not yet well integrated, the markers from the two maps are represented separately. Clearly, integration of the two maps and addition of new markers, with the identification of their corresponding YAC clones, will result in a high proportion of this chromosome arm being contained in mapped YAC clones.

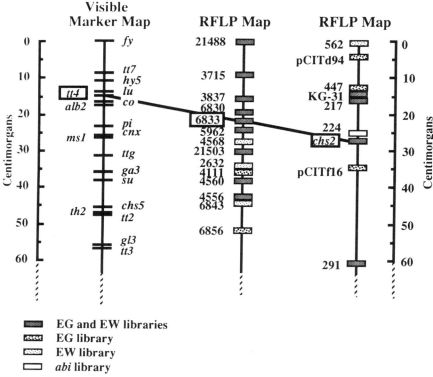

Figure 5 YAC clone coverage on the top half of chromosome 5. YAC contigs corresponding to RFLP markers are represented as boxes in the map position that has been established for the individual RFLP markers. Different fillings of the boxes indicate if YAC clones could be found in the EG and the EW libraries or only in one. The conversion of 1 cM for 140 kb was used to convert the size of the YAC contigs into genetic distance (Chang et al. 1988). Each YAC contig spans on average 240 kb of DNA (Hwang et al. 1991). The two published RFLP maps (Chang et al. 1988; Nam et al. 1989) have been represented along the visible marker map (Koornneef 1990). The identified contact point between the three maps (tt4, 6833, chs2) is indicated by a line.

Linking up YAC clones hybridizing to adjacent RFLP markers

Linking of the YAC clones hybridizing to adjacent RFLP markers is being achieved by chromosome walking experiments. This involves the isolation of end probes from one YAC clone and hybridization of these probes to the YAC libraries to identify overlapping YAC clones. The generation of end probes has been achieved in this laboratory using two strategies (both of which are summarized in Fig. 6): inverse polymerase chain reaction (IPCR) (Ochman et al. 1988) and plasmid rescue in *E. coli* of the left end of a YAC. Total yeast DNA is used as the starting material for both techniques.

For IPCR experiments, a variety of restriction enzymes are used to digest the yeast DNA to appropriate fragment sizes. The DNA is diluted to promote circle formation of the generated restriction fragments in the ligation reaction. The oligonucleotides are added and the PCR reactions are performed. *Alu*I can be used to generate fragments to form circles for the left and the right end of the YAC inserts. *Eco*RV, *Cla*I, and *Sac*I have been used successfully for the left end of the YAC inserts and *Sal*I, *Hinc*II, and *Nae*I for the right end. Figure 6 schematically represents the ligated DNA molecules generated from the right end of a YAC insert after utilizing the enzymes *Alu*I and *Hinc*II. Generally, it was necessary to test more than one enzyme for each of the YAC insert ends to generate a PCR product of sufficient size for use as a probe.

The alternative technique of plasmid rescue takes advantage of the fact that the left end of a YAC, using the YAC vectors described by Burke et al. (1987), contains an origin of replication and an ampicillin resistance gene, both of which are functional in *E. coli*. Total yeast DNA is digested with either *Xho*I or *Nde*I and subsequently ligated under dilute conditions. The generated circles (Fig. 6) are then transformed into *E. coli* cells. Of all the circles generated, only the one containing the left end vector sequences and the adjacent plant DNA can replicate in *E. coli* and express ampicillin resistance and, thus, can be easily selected for. With the YAC vectors used for the Arabidopsis libraries (pYAC3, pYAC4, and pYAC41), only fragments adjacent to the left end of the YAC

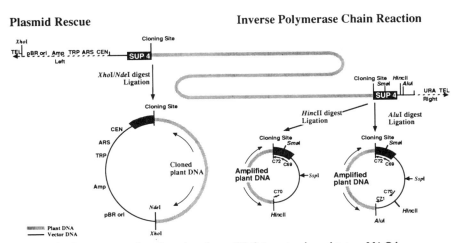

Figure 6 Generation of end probes from YAC inserts cloned into pYAC4 or a derivative (Burke et al. 1987). The left part of the figure shows the structure of the circle, which is cloned in plasmid rescue experiments. The right part of the figure depicts the details of the circles that are used in combination with appropriate oligonucleotides (e.g., C69 and C70) in PCR experiments to amplify the YAC insert sequences adjacent to the YAC vector sequences. The figure shows circles derived from the right end of the YAC insert.

insert can be cloned by using plasmid rescue experiments unless an origin of replication and a suitable resistance marker are introduced by homologous recombination into the right end of the YAC clones. YAC end fragments generated by the method of plasmid rescue are longer than the end fragments that are generated in IPCR experiments. This has the advantage that small stretches of repeated sequence that may lie at the ends of the insert, and make the IPCR-generated probes useless, can be removed. To use the end fragments as probes in colony hybridization experiments, all vector sequences need to be removed by using appropriate restriction enzyme digests and subsequent gel purification of the plant DNA insert fragment.

We have hybridized 64 probes derived from the ends of the YAC inserts to the YAC libraries. The majority of the end probes hybridized to a small number of colonies. Eleven of them, however, hybridized to many YAC clones, indicating that the end probe contained a stretch of repetitive DNA. Our walking experiments have concentrated on a 10-cM chromosomal region on chromosome 4 lying between RFLP markers 210 and 226 (Chang et al. 1988). We now have two contiguous regions of 915 and 1300 kb within this region, involving 84 YAC clones. Analysis of these contigs has shown that at least 8 of the YAC clones involved are chimeric. End probes of these clones did not hybridize to YAC clones, which should be adjacent or which span the region. The problem of chimeric YAC clones has also been described for human YAC libraries. They can represent a very high proportion (40–60%) of the library (compare with Green et al. 1991), and their identification is laborious and time-consuming. Our preliminary estimates for the libraries investigated so far (EG and EW), based on a detailed analysis of 59 YAC clones, range between 10% and 20% of the YAC clones. This level should not be a problem with chromosome walking experiments. It does mean, however, that it is vital to achieve duplicate YAC coverage for each position in the genome. The use of new libraries without chimeric YAC clones would significantly reduce the number of YAC clones required to ensure complete genome coverage. This has been demonstrated by the calculations of Arratia et al. (1991).

Hybridization of repeat sequences to the Arabidopsis YAC libraries

Despite the higher representation of genomic sequences in the YAC clones as compared to the cosmid clones, there may still be chromosomal regions, notably those carrying highly repeated DNA, e.g., from centromeric and sub-telomeric regions, that could prove difficult to span even using YAC contigs. These regions are currently the specific focus of two groups (Richards and Ausubel 1988; Richards et al. 1991; J.

Ecker, pers. comm.). We decided to identify the YAC clones carrying characterized repeat sequences in order to monitor at which point the chromosomal walks run into regions of repeated DNA. There are relatively few repeat sequences within total Arabidopsis DNA (Leutwiler et al. 1984; Pruitt and Meyerowitz 1986). The main classes of repeat sequences are chloroplast DNA and rDNA (Pruitt and Meyerowitz 1986). In addition, a highly repeated DNA sequence of 180 bp has been described which is arranged in tandem arrays and constitutes approximately 1–1.5% of the genome (Martinez-Zapater et al. 1986; Simoens et al. 1988). It is also closely related to a 500-bp repeat sequence described by Simoens et al. (1988).

The YAC libraries were constructed from total plant DNA, so it is expected that some of the YAC clones will contain chloroplast DNA. When the EW YAC library was hybridized with a mixture of cosmid clones carrying chloroplast DNA (gift of A. Brennicke and W. Schuster, IGF, Berlin), approximately 7% of the colonies showed a positively hybridizing signal, although to differing strengths (R. Schmidt et al., unpubl.). These clones have not been discarded from subsequent screening procedures, because chloroplast DNA-related sequences have been shown to be present in the nuclear genome (Timmis and Steele Scott 1985). Differing levels of homology of these nuclear sequences to the chloroplast DNA would explain why some YAC clones show only faint hybridization signals to some chloroplast DNA probes. The coordinates of the YAC clones have been recorded (R. Schmidt et al., in prep.). If one of these YAC clones which shows a strong hybridization signal to chloroplast DNA also hybridizes to probes generated from chromosome walks or to RFLP markers, it is investigated further because it possibly represents a chimeric YAC clone.

The rDNA is present in 570 copies and represents about 8% of the genome (Pruitt and Meyerowitz 1986). An rDNA probe has been used in in situ hybridization experiments to localize the rDNA loci on the Arabidopsis chromosomes 2 and 4 (Maluszynska and Heslop-Harrison 1991). YAC clones from the different YAC libraries carrying rDNA sequences are currently being characterized.

The in situ hybridization analysis of the 180-bp repeat sequence revealed its colocalization with the centromeric heterochromatin of all five Arabidopsis chromosomes (Maluszynska and Heslop-Harrison 1991). This sequence hybridized strongly to many YAC clones in the colony hybridization experiments. Approximately 10% of the EW clones showed hybridization to this repeat sequence, whereas only 2% did so in the EG YAC library. The 500-bp repeat yielded an almost identical hybridization pattern to the 180-bp repeat in the colony hybridization experiments (Fig. 7). Because the 180-bp sequence and the 500-bp sequence together represent approximately 2% of the genome (Simoens et al. 1988), and the EG and EW libraries each contain about 3 genome equivalents, it is clear that these repeat families are underrepresented in the EG library. A probable reason for this is an unequal distribution of *Bam*HI sites in the tandem arrays of the repeat sequences. The coordinates of the YAC clones hybridizing to the repeat sequences have been established for the EG and EW library (R. Schmidt et al., in prep.). If the end probes from chromosome walks hybridize to these YAC clones, it is likely that the walks have reached the centromeric heterochromatin region.

A relatively high level of instability of YAC clones carrying these repeated DNA sequences has been found. Approximately 20 of the 100 clones analyzed showed multiple YACs, all of which are hybridizing to the repeat DNA probes. Neil et al. (1990) also observed instability of YACs containing human tandemly repeated sequences. Some instability has also been detected with YACs hybridizing to various RFLP markers. Among the 187 yeast colonies we have analyzed, 5 were shown to contain two YACs, both of which hybridized to the corresponding RFLP marker DNA. We believe the smaller YAC is a stable deletion product from the larger one. Whether this instability is also associated with repeats is being examined. Similar observations have been made by Abidi et al. (1990) for YACs containing human DNA sequences.

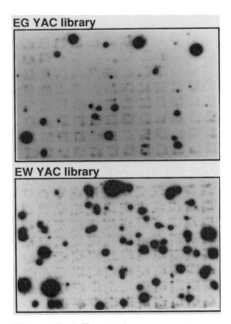

Figure 7 Differential representation of a repeat sequence in YAC libraries. Hybridization of the 500-bp repeat sequence (Simoens et al. 1988) to filters carrying 768 colonies (8 offsets) of the EG (Grill and Somerville 1991a) library and the EW (Ward and Jen 1990) library.

STRATEGIES FOR COMPLETING THE PHYSICAL MAP

Techniques used in physical mapping in other organisms not available in Arabidopsis

Arabidopsis meets many criteria for a model organism, but it lacks some features that would ease the establishment and completion of a physical map. One is that in situ hybridization with single-copy sequences has not been reported with Arabidopsis metaphase chromosomes. Even when successful, the very small size of the Arabidopsis chromosomes would mean that elegant cytological mapping experiments described, for example, with *Drosophila* (Garza et al. 1989) or human YACs (Driesen et al. 1991) are unlikely to be feasible. The small size of the chromosomes and their relatively homogeneous length also means that flow-sorting of Arabidopsis chromosomes is unlikely to be useful in the generation of chromosome-specific libraries. Furthermore, hybrids of Arabidopsis with other plants have not been described in enough detail to parallel the well-studied human-hamster cell lines, which have been so important in the construction of human YAC libraries (see, e.g., Zucchi and Schlessinger 1992).

Capitalizing on the advantages of Arabidopsis

Strategies for completing the physical map need to capitalize on the main advantages of the Arabidopsis genome; namely, its extremely small size, the presence of relatively few repeated DNA sequences, and the increasing number of markers being mapped onto the genome. The first two of these parameters have been and will continue to be crucial in the construction and linking of the cosmid contigs. In terms of completing the overlapping YAC library, hybridization of 125 markers has already generated YAC contigs covering approximately 30% of the genome (Hwang et al. 1991). Furthermore, even with this relatively small number of markers, YAC contigs overlapping 2 or more RFLP markers have been identified (Grill and Somerville 1991a; Hwang et al. 1991). Major advances in the construction of overlapping YAC clones will be the availability of three resources: larger YAC libraries, libraries carrying fewer chimeric YAC clones, and additional mapped markers.

The first of these requirements has been facilitated by the availability of the YAC library described by Ecker (1990), which has an average insert size of 250 kb. With this library, Ewens et al. (1991) have calculated that by using 500 probes, 87% of the Arabidopsis genome could be covered in 180 contigs of an average size of 550 kb. This library was also constructed with special attention paid to minimizing the number of chimeric clones. This should significantly reduce the number of YAC clones required to ensure complete genome coverage (Arratia et al. 1991).

The number of RFLP and RAPD markers mapped onto the Arabidopsis genetic map is increasing rapidly. Unfortunately, there are at least three mapping populations being used with a relatively small number of common markers used on each population. Thus, the integration of the maps is rather poor. A statistical integration using the raw scoring data is being attempted (P. Stam and M. Koornneef, pers. comm.). In the long term, this should not be a problem, as several groups have generated recombinant inbred lines by a process illustrated in Figure 8. Two inbred parents are crossed, and the progeny is self-pollinated to provide the F_2 generation. Individual F_2 plants are then selfed again to produce the F_3 generation, and the process is continued

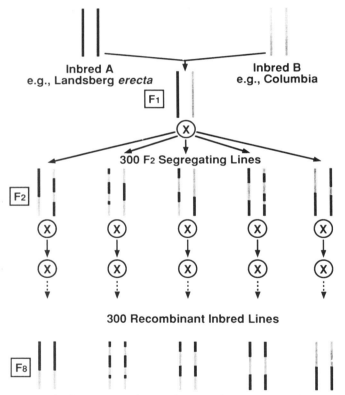

Figure 8 Generation of recombinant inbreds. Plants of two Arabidopsis ecotypes (e.g., Landsberg *erecta* and Columbia) are crossed. The F_1 is allowed to self-pollinate and many individual F_2 lines are generated. Approximately 300 segregating F_2 lines are randomly chosen. One seed of each of the lines, following self-pollination, is used to generate the F_3 plants. This process of self-pollination of single progeny plants from each of the lines is continued to the F_8 generation. At this stage, there are 300 recombinant inbred lines, each containing fixed short linkage blocks of DNA derived from one of the two parents. Propagation by selfing will then provide an inexhaustible supply of seed for mapping DNA fragments onto the Arabidopsis genome.

by single-seed descent until the F_8 generation to generate homozygous lines. By F_8 there is theoretically only a 0.78% chance of the lines still being heterozygous at a particular locus. The F_9 seed from the individual lines and further generations are then bulk harvested to provide an inexhaustible supply of seed for distribution throughout the Arabidopsis community for mapping purposes. Previous mapping populations have used F_3 or F_4 families, which, once exhausted, have to be reconstructed from the beginning. Once the previously mapped probes have also been mapped onto these lines, new probes can then be added by any laboratory and the mapping information can be directly compared. One of the recombinant inbred populations has the Arabidopsis ecotypes Columbia and Landsberg *erecta* as parents in the original cross (C. Lister and C. Dean, in prep.). The other used the ecotype WS (Wassilewskija) and the mapping line W100 (Koornneef et al. 1987), which is a Landsberg *erecta* line carrying 10 visible markers, one on each chromosome arm (Reiter et al. 1992).

The recombinant inbred lines can be used to target markers to a specific region of the genome, as demonstrated by Reiter et al. (1992), by modifying a technique that has been introduced by Michelmore et al. (1991). They have shown that markers linked to a specific gene or genomic region can be rapidly identified by a procedure termed bulked segregant analysis. This technique relies on two bulked DNA samples generated using a polymorphic cross from a segregating population for a particular locus. A bulk contains individuals identical for the particular trait, but arbitrary at unlinked regions. RFLP probes or RAPD primers can be used to identify polymorphisms, linked to the locus of interest, between the two bulks, because they are only differing in the region of the selected locus. Markers were reliably identified in a 25-cM window at either side of the locus. Reiter et al. (1992) created bulks of recombinant inbreds based on their genotype and used the resulting samples successfully to select for markers located on chromosome 1.

Increasing the number of markers on the Arabidopsis genetic map

A large number of clones derived from many different types of experiments are likely to be mapped once the recombinant inbred lines have been widely distributed. This will include the mapping of a large number of cDNA clones isolated from different tissue- and stage-specific libraries and sequences homologous to newly cloned genes. In addition, a significant number of new markers will be added to the Arabidopsis RFLP map through the development of an insertional mutagenesis system in Arabidopsis.

Two types of mutagenic elements have been employed for insertional mutagenesis experiments (schematically illustrated in Fig. 9). The

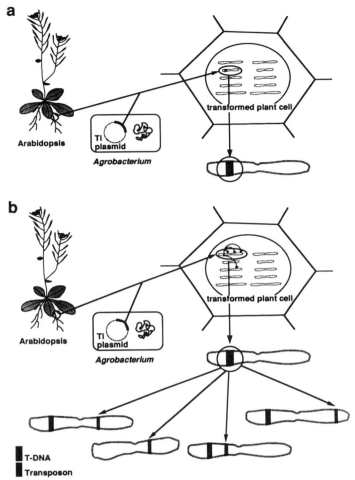

Figure 9 Random insertions of T-DNA and heterologous transposable elements into the genome of Arabidopsis. (*a*) Random integration of the T-DNA into the Arabidopsis genome after infection of plant cells with *Agrobacterium tumefaciens*. (*b*) Random integration of a T-DNA containing a transposable element. Upon excision and reinsertion of the transposon, new random integration sites in either the same or a different chromosome are obtained. The transposition events are independently happening in many plant cells. Four different examples are shown. In three cases, the transposon reinserted into the same chromosome as the T-DNA; in one case, the integration is demonstrated for a different chromosome.

first is the T-DNA from the Ti plasmid of *A. tumefaciens*. More than 10,000 lines carrying an introduced T-DNA insertion have been generated from a number of laboratories, many of which result in a mutation (see, e.g., Feldmann 1991). Mapping of the different T-DNA insertions, starting with those giving a mutant phenotype, will add a large number of new markers to the Arabidopsis map. The second class of insertional mutagen being developed is a transposable element. Because Arabidopsis does not have a well-characterized endogenous transposon system, a number of heterologous transposons from either maize or snapdragon (for review, see Bhatt and Dean 1992) have been introduced into Arabidopsis using *Agrobacterium* transformation. Transposons have the advantage over T-DNA tagging in that they continue to excise and reintegrate after the transformation step and thus can continue to cause new mutations. In addition, if a two-element system is used, consisting of a nonautonomous element and a nonmobile transposase source, any mutations caused by the tissue culture steps used in the initial transformation can be discarded before the two components of the transposon system are brought together.

In Arabidopsis, the maize transposons *Ac* (*Activator*) and *Ds* (*Dissociation*) appear to be the most useful heterologous transposons. These elements preferentially transpose to linked sites in the genome. It is beneficial, therefore, in a tagging experiment, which is targeting a particular locus, to start with a line carrying a transposon in a nearby position in the genome. For this reason, a large number of T-DNA insertions carrying various modifications of the *Ac* and *Ds* elements are being mapped onto the Arabidopsis RFLP map. IPCR is used to amplify the flanking plant DNA. The IPCR fragment is then cloned and used to detect polymorphisms between the parents used to generate the recombinant inbred lines, namely Landsberg *erecta* and Columbia. The fragment is then mapped onto the recombinant inbred lines. Our aim is to map at least 50 T-DNA insertions so that we can start tagging experiments from linked sites anywhere in the genome. These markers are also being used against the YAC libraries (C. Recknagel and G. Coupland, pers. comm.; C. Lister and C. Dean, unpubl.).

Sequencing of the Arabidopsis genome

The final aim of the Arabidopsis physical mapping project will be the complete sequence analysis of the nuclear and organellar genomes. The sequencing of the chloroplast and mitochondrial genomes has been initiated (Meyerowitz et al. 1991), and 160 kb of unique sequence in numerous small linkage groups has already been established for the mitochondrial genome (Grönger et al. 1991). Large-scale sequencing of the nuclear genome has already been started by H. Goodman, and two 40-kb cosmid clones have been sequenced (Hauge et al. 1991). There are also several international programs to fund large-scale sequencing of genomic regions and random cDNA clones (Meyerowitz et al. 1991). Databases to store all the information produced from the mapping and the sequence analysis are being established (Meyerowitz et al. 1991).

This concentration of effort and easy access to the ever-increasing amount of information on the Arabidopsis genome should soon ensure that gene cloning is no longer a major limiting factor in the dissection of complex plant processes.

Acknowledgments

We thank D. Weigel and E. Meyerowitz (Pasadena) for their help with the plasmid rescue; E. Meyerowitz, B. Hauge, and H. Goodman (Boston) for providing RFLP markers prior to publication; and A. Brennicke and W. Schuster (Berlin) for the chloroplast DNA clones. David Flanders is gratefully acknowledged for help with some of the figures. We thank Jo Putterill, Prisca Stabel, Anuj Bhatt, David Flanders, Graham Moore, and Brian Staskawicz for critically reading the manuscript.

References

Abidi, F.E., M. Wada, R.D. Little, and D. Schlessinger. 1990. Yeast artificial chromosomes containing human Xq24-Xq28 DNA: Library construction and representation of probe sequences. *Genomics* 7: 363.

Araki, T., S. Naito, and Y. Komeda. 1989. Characterization of the late flowering mutant *gi2* of *Arabidopsis thaliana*. In *The genetics and molecular biology of* Arabidopsis, Abstract no. II,9. Indiana University, Bloomington, Indiana.

Arratia, R., E.S. Lander, S. Tavaré, and M.S. Waterman. 1991. Genomic mapping by anchoring random clones: A mathematical analysis. *Genomics* 11: 806.

Arumuganathan, K. and E.D. Earle. 1991. Nuclear DNA content of some important plant species. *Plant Mol. Biol. Rep.* 9: 208.

Bhatt, A.M. and C. Dean. 1992. Development of tagging systems in plants using heterologous transposons. *Curr. Opin. Biotech.* 3: 152.

Bleecker, A. 1991. Genetic analysis of ethylene responses in *Arabidopsis thaliana*. In *Molecular biology of plant development* (ed. G.I. Jenkins and W. Schuch), p. 149. Company of Biologists, Cambridge, England.

Burke, D.T., G.F. Carle, and M.V. Olson. 1987. Cloning of large segments of exogenous DNA into yeast by means of artificial chromosome vectors. *Science* 236: 806.

Chang, C., J.L. Bowman, A.W. DeJohn, E.S. Lander, and E.M. Meyerowitz. 1988. Restriction fragment length polymorphism linkage map for *Arabidopsis thaliana*. *Proc. Natl. Acad. Sci.* 85: 6856.

Chaudhury, A.M., S. Craig, K. Blomer, L. Farrell, R. Chapple, B. Sherman, and L. Dennis. 1991. Genetic control of male-fertility in *Arabidopsis thaliana*. In *Robertson Symposium on* Arabidopsis thaliana *and the molecular basis of plant biology*. Canberra, Australia. [Abstract.]

Coulson, A., J. Sulston, S. Brenner, and J. Karn. 1986. Toward a physical map of the genome of the nematode *Caenorhabditis elegans*. *Proc. Natl. Acad. Sci.* 83: 7821.

Coulson, A., R. Waterston, J. Kiff, J. Sulston, and Y. Kohara. 1988. Genome linking with yeast artificial chromosomes. *Nature* **335**: 184.

Coulson, A., Y. Kozono, B. Lutterbach, R. Shownkeen, J. Sulston, and R. Waterston. 1991. YACs and the *C. elegans* genome. *BioEssays* **13**: 413.

Cross, S.H. and P.F.R. Little. 1986. A cosmid vector for systematic chromosome walking. *Gene* **49**: 9.

Cruz-Alvarez, M., J.A. Jarillo, A. Leyva, and J.M. Martinez-Zapater. 1991. Characterization of late flowering mutations at the *fve* locus of Arabidopsis. In *Third International Congress of Plant Molecular Biology: Molecular biology of plant growth and development*. Tucson, Arizona. [Abstract no. 494.]

Debener, T., H. Lehnackers, M. Arnold, and J.L. Dangl. 1991. Identification and molecular mapping of a single *Arabidopsis thaliana* locus determining resistance to a phytopathogenic *Pseudomonas syringae* isolate. *Plant J.* **1**: 289.

Delaney, T.P., A.E. Pepper, and J. Chory. 1991. Progress toward cloning the *de-etiolated-1* (*det-1*) locus in *Arabidopsis thaliana*. In *Third International Congress of Plant Molecular Biology: Molecular biology of plant growth and development*. Tucson, Arizona. [Abstract no. 1476.]

Driesen, M.S., J.G. Dauwerse, M.C. Wapenaar, E.J. Meershoek, P. Mollevanger, K.L. Chen, K.H. Fishbeck, and G.J.B. van Ommen. 1991. Generation and fluorescent *in situ* hybridization mapping of yeast artificial chromosomes of 1p, 17p, 17q, and 19q from a hybrid cell line by high-density screening of an amplified library. *Genomics* **11**: 1079.

Ecker, J.R. 1990. PFGE and YAC analysis of the Arabidopsis genome. *Methods* **1**: 186.

Ewens, W.J., C.J. Bell, P.J. Donnelly, P. Dunn, E. Matallana, and J.R. Ecker. 1991. Genome mapping with anchored clones: Theoretical aspects. *Genomics* **11**: 799.

Feinbaum, R.L. and F.M. Ausubel. 1988. Transcriptional regulation of the *Arabidopsis thaliana* chalcone synthase gene. *Mol. Cell. Biol.* **8**: 1985.

Feldmann, K.A. 1991. T-DNA insertion mutagenesis in Arabidopsis: Mutational spectrum. *Plant J.* **1**: 71.

Feldmann, K.A. and M.D. Marks. 1987. *Agrobacterium*-mediated transformation of germinating seeds of *Arabidopsis thaliana*: A non-tissue culture approach. *Mol. Gen. Genet.* **208**: 1.

Garza, D., J.W. Ajioka, D.T. Burke, and D.L. Hartl. 1989. Mapping the *Drosophila* genome with yeast artificial chromosomes. *Science* **246**: 641.

Green, E.D., H.C. Riethman, J.E. Dutchik, and M.V. Olson. 1991. Detection and characterization of chimeric yeast artificial-chromosome clones. *Genomics* **11**: 658.

Grill, E. and C. Somerville. 1991a Construction and characterization of a yeast artificial chromosome library of Arabidopsis which is suitable for chromosome walking. *Mol. Gen. Genet.* **226**: 484.

———. 1991b. Development of a system for efficient chromosome walking in Arabidopsis. In *Molecular biology of plant development* (ed. G.I. Jenkins and W. Schuch), p. 57. Company of Biologists, Cambridge, England.

Grönger, P., M. Unseld, U. Eckert, and A. Brennicke. 1991. Sequence analysis of a mitochondrial genome of *Arabidopsis thaliana*—A progress report. In *Third International Congress of Plant Molecular Biology: Molecular biology of plant growth and development*. Tucson, Arizona. [Abstract no. 2050.]

Hauge, B.M., S. Hanley, J. Giraudat, and H.M. Goodman. 1991. Mapping the Arabidopsis genome. In *Molecular biology of plant development* (ed G.I. Jenkins and W. Schuch), p. 45. Company of Biologists, Cambridge, England.

Hwang, I., T. Kohchi, B.M. Hauge, H.M. Goodman, R. Schmidt, G. Cnops, C. Dean, S. Gibson, K. Iba, B. Lemieux, V. Arondel, L. Danhoff, and C. Somerville. 1991. Identification and map position of YAC clones comprising one-third of the Arabidopsis genome. *Plant J.* 1: 367.

Koornneef, M. 1990. Linkage map of *Arabidopsis thaliana* 2N = 10. In *Genetic maps, Locus maps of complex genomes, Book 6, Plants* (ed. S.J. O'Brien), p. 6.94. Cold Spring Harbor Laboratory Press, New York.

Koornneef, M., C.J. Hanhardt, E.P. van Loenen-Martinet, and J.H. van der Veen. 1987. A marker line, that allows the detection of linkage on all Arabidopsis chromosomes. *Arabidopsis Inf. Serv.* 23: 46.

Koornneef, M., C.J. Hanhart, E.P. van Loenen Martinet, A.J.M. Peeters, and J.H. van der Veen. 1991. Genes controlling flowering time in *Arabidopsis thaliana*. In *Third International Congress of Plant Molecular Biology: Molecular biology of plant growth and development*. Tucson, Arizona. [Abstract no. 509.]

Koornneef, M., J. van Eden, C.J. Hanhart, P. Stam, F.J. Braaksma, and F.J. Feenstra. 1983. Linkage map of *Arabidopsis thaliana. J. Hered.* 74: 265.

Lemieux, B., B. Hauge, and C. Somerville. 1991. RFLP mapping and chromosome walking to the *Arabidopsis thaliana fae1* locus. In *Third International Congress of Plant Molecular Biology: Molecular biology of plant growth and development*. Tucson, Arizona. [Abstract no. 727.]

Leung, J., N. Vartanian, B. Hauge, H.M. Goodman, and J. Giraudat. 1991. Molecular genetic analysis of abscisic acid action in *Arabidopsis thaliana*. In *Robertson Symposium on* Arabidopsis thaliana *and the molecular basis of plant biology*. Canberra, Australia. [Abstract.]

Leutwiler, L.S., B.R. Hough-Evans, and E.M. Meyerowitz. 1984. The DNA of *Arabidopsis thaliana. Mol. Gen. Genet.* 194: 15.

Leyser, O., C. Lincoln, J. Turner, D. Lammer, and M. Estelle. 1991. Molecular characterisation of the *axr1* locus of *Arabidopsis thaliana*. In *Third International Congress of Plant Molecular Biology: Molecular biology of plant growth and development*. Tucson, Arizona. [Abstract no. 863.]

Maluszynska, J. and J.S. Heslop-Harrison. 1991. Localization of tandemly repeated DNA sequences in *Arabidopsis thaliana. Plant J.* 1: 159.

Martinez-Zapater, J.M., M.A. Estelle, and C.R Somerville. 1986. A highly repeated DNA sequence in *Arabidopsis thaliana. Mol. Gen. Genet.* 204: 417.

Mayer, U., R.A. Torres Ruiz, T. Berleth, S. Miséra, and G. Jürgens. 1991. Mutations affecting body organization in the Arabidopsis embryo. *Nature* 353: 402.

Medd, J., O. Dolezal, and C. Cobbett. 1991. Arabinose metabolism mutants of *Arabidopsis thaliana*. In *Robertson Symposium on* Arabidopsis thaliana *and the molecular basis of plant biology*. Canberra, Australia. [Abstract.]

Meyerowitz, E., C. Dean, R. Flavell, H. Goodman, M. Koornneef, J. Peacock, Y. Shimura, C. Somerville, and M. van Montagu. 1991. *The multinational coordinated* Arabidopsis thaliana *genome research project. Progress report: Year one.* Publication no. 91-60. National Science Foundation, Washington, D.C.

Michelmore, R.W., I. Paran, and R.V. Kesseli. 1991. Identification of markers linked to disease-resistance genes by bulked segregant analysis: A rapid method to detect markers in specific genomic regions using segregating populations. *Proc. Natl. Acad. Sci.* **88**: 9828.

Nagpal, P. and J. Chory. 1991. Progress toward cloning the Arabidopsis *det2* gene. In *Third International Congress of Plant Molecular Biology: Molecular biology of plant growth and development.* Tucson, Arizona. [Abstract no. 1485.]

Nam, H.-G., J. Giraudat, B. den Boer, F. Moonan, W.D.B. Loos, B.M. Hauge, and H. Goodman. 1989. Restriction fragment length polymorphism linkage map of *Arabidopsis thaliana. Plant Cell* **1**: 699.

Neil, D.L., A. Villasante, R.B. Fisher, D. Vetrie, B. Cox, and C. Tyler-Smith. 1990. Structural instability of human repeated DNA sequences cloned in yeast artificial chromosome vectors. *Nucleic Acids Res.* **18**: 1421.

Ochman, H., A.S. Gerber, and D.L. Hartl. 1988. Genetic application of an inverse polymerase chain reaction. *Genetics* **120**: 621.

Pruitt, R.E. and E.M. Meyerowitz. 1986. Characterization of the genome of *Arabidopsis thaliana. J. Mol. Biol.* **187**: 169.

Putterill, J., F. Robson, K. Ingle, and G. Coupland. 1991. Towards isolation of the Arabidopsis flowering-time gene *fg.* In *Third International Congress of Plant Molecular Biology: Molecular biology of plant growth and development.* Tucson, Arizona. [Abstract no. 496.]

Reiter, R.S., J.G.K. Williams, K.A. Feldmann, J.A. Rafalski, S.V. Tingey, and P.A. Scolnik. 1992. Global and local genome mapping in *Arabidopsis thaliana* by using recombinant inbred lines and random amplified polymorphic DNAs. *Proc. Natl. Acad. Sci.* **89**: 1477.

Richards, E.J. and F.M. Ausubel. 1988. Isolation of a higher eukaryotic telomere from *Arabidopsis thaliana. Cell* **53**: 127.

Richards, E.J., H.M. Goodman, and F.M. Ausubel. 1991. The centromere region of *Arabidopsis thaliana* chromosome 1 contains telomere-similar sequences. *Nucleic Acids Res.* **19**: 3351.

Simoens, C.R., J. Gielen, M. Van Montagu, and D. Inzé. 1988. Characterization of highly repetitive sequences of *Arabidopsis thaliana. Nucleic Acid Res.* **16**: 6753.

Sulston, J., F. Mallett, R. Durbin, and T. Horsnell. 1989. Image analysis of restriction enzyme fingerprint autoradiograms. *Cabios* **5**: 101.

Sulston, J., F. Mallett, R. Staden, R. Durbin, T. Horsnell, and A. Coulson. 1988. Software for genome mapping by fingerprinting techniques. *Comput. Appl. Biosci.* **4**: 125.

Timmis, J.N. and N. Steele Scott. 1985. Movement of genetic information between the chloroplast and nucleus. In *Genetic flux in plants* (ed. B. Hohn and E.S. Dennis), p. 61. Springer-Verlag, New York.

Valvekens, D., M. van Montagu, and M. van Lijsebettens. 1988. *Agrobacterium tumefaciens*-mediated transformation of Arabidopsis root explants using kanamycin selection. *Proc. Natl. Acad. Sci.* **85**: 5536.

Walker, A.R. and J.C. Gray. 1991. Trichome development in Arabidopsis: Cloning of *ttg.* In *Robertson Symposium on* Arabidopsis thaliana *and the molecular basis of plant biology.* Canberra, Australia. [Abstract.]

Ward, E.R. and G.C. Jen. 1990. Isolation of single-copy-sequence clones from a yeast artificial chromosome library of randomly-sheared *Arabidopsis thaliana* DNA. *Plant Mol. Biol.* **14:** 561.

Westphal, L., R.M. Ewing, I. Bancroft, and C. Dean. 1991. Cloning *fca*, a late-flowering locus of *Arabidopsis thaliana* (L.) Heynh. In *Third International Congress of Plant Molecular Biology: Molecular biology of plant growth and development*. Tucson, Arizona. [Abstract no. 508.]

Williams, J.G.K., A.R. Kubelik, K.J. Livak, J.A. Rafalski, and S.V. Tingey. 1990. DNA polymorphisms amplified by arbitrary primers are useful as genetic markers. *Nucleic Acids Res.* **18:** 6531.

Wilson, A., F.B. Pickett, J. Turner, and M. Estelle. 1990. Isolation and characterization of a dominant hormone-resistant mutant of Arabidopsis. In *Fourth International Conference on Arabidopsis Research*. Vienna, Austria. [Abstract no. 140.]

Wilson, Z.A., J. Dawson, M.G.M. Aarts, P. Anthony, L.G. Briarty, and B.J. Mulligan. 1991. Genetic male sterility in *Arabidopsis thaliana*. In *Robertson Symposium on* Arabidopsis thaliana *and the molecular basis of plant biology*. Canberra, Australia. [Abstract.]

Yanofsky, M.F., H. Ma, J.L. Bowman, G.D. Drews, K.A. Feldmann, and E.M. Meyerowitz. 1990. The protein encoded by the Arabidopsis homeotic gene *agamous* resembles transcription factors. *Nature* **346:** 35.

Zucchi, I. and D. Schlessinger. 1992. Distribution of moderately repetitive sequences pTR5 and LF1 in Xq24-q28 human DNA and their use in assembling YAC contigs. *Genomics* **12:** 264.

Porcine Genome Analysis

Chris S. Haley and Alan L. Archibald

AFRC Institute of Animal Physiology and Genetics Research
Edinburgh Research Station, Roslin
Midlothian, EH25 9PS, Scotland

The mapping of the porcine genome is proceeding using the same technologies and methods that have been developed and applied to the analysis of other genomes. A major reason for developing a porcine map is to provide a tool for the detection and study of genes controlling quantitative traits (quantitative trait loci or QTL) through their linkage to marker loci. The map will also allow the selection of favorable QTL alleles via selection on linked markers (marker-assisted selection). Both the production of the marker linkage map and the detection of QTL are facilitated by the use of a cross between genetically divergent breeds. F_2 intercrosses of European commercial breeds with the Chinese Meishan or the European wild boar are being used to provide reference families for mapping. Probes based on homologous or heterologous sequences, mostly cDNA sequences, can be used to detect restriction-fragment-length polymorphisms (RFLPs), which are highly informative in these intercrosses. Such probes are being used to provide a skeleton map that can ultimately be directly aligned with maps of other mammals, particularly mouse and humans. The map is being fleshed out with variable number of tandem repeat (VNTR)-based markers, especially those with a microsatellite core sequence, which will be informative both between and within breeds. Physical assignment of some markers is being achieved through the use of in situ hybridization, hybrid cell lines, and flow-sorted chromosomes; these are facilitated by the polydisperse character of the porcine karyotype. In Europe, coordination of porcine genome analysis is under the umbrella of the Pig Gene Mapping Project, which has been set up under the BRIDGE scheme of the Commission of the European Communities and which provides a framework for collaboration, enabling the map to be efficiently produced.

Genome Analysis Volume 4: *Strategies for Physical Mapping*
©1992 Cold Spring Harbor Laboratory Press 0-87969-412-2/92 $3 + 00

The main points are:

❑ Mapping provides a tool for the study and manipulation of QTL in farm species and could allow development of improved selection methods.

❑ Genetically divergent breeds are available in the pig, and F_2 intercrosses between breeds provide an ideal resource both for production of the marker linkage map and for mapping QTL.

❑ RFLP markers based on cDNA probes can be highly informative in a breed cross and provide a means of linking the porcine map directly to the human and murine maps, whereas VNTR-based markers provide a means of fleshing out the map so that it can be used both between and within breeds.

❑ Physical mapping is facilitated by the polydisperse porcine karyotype, which eases the use of in situ hybridization and allows development of a flow-sorted karyotype.

❑ Cloning of mapped QTL will be difficult but will likely be via the candidate gene approach and facilitated by the alignment of the porcine and human maps. The halothane locus provides a paradigm for such an approach.

INTRODUCTION

Man has effected the genetic improvement of plant crop and domestic animal species over the preceding millennia with minimal knowledge of the individual actions of genes that contribute to phenotypic variation. This "black box" approach has been effective because phenotypic selection is relatively robust to failings in the genetic model used. It is usually assumed that there are an infinite number of unlinked loci each of infinitesimal effect that combine to control variation in the traits under selection (hence, the "infinitesimal model"). Although this model is obviously unrealistic, it is reasonably effective for the prediction of short-term selection responses, even if it is inadequate in the longer term (see, e.g., Meyer and Hill 1991). It would be very valuable to have a better understanding of the genetic architecture of quantitative variation.

A knowledge of the sizes, locations, actions, and interactions of the major loci contributing to quantitative variation (quantitative trait loci or QTL) would allow better statistical models to be built for use in breeding programs. Identification of the individual loci responsible would provide a better understanding of the biological basis of quantitative genetic variation and open the way to selection directly on the genotype, rather than on the effects of the genotype as mirrored in the phenotype. Only QTL with very large effects can be detected directly through their influence on the phenotype (e.g., through segregation analysis; Hill and Knott 1990), but QTL of much more modest effect can be detected via linkage to polymorphic genetic markers. Thus, a genetic map can provide a starting point for understanding the genetic control of performance. Because the language of the genome is universal, the techniques and approaches being deployed to study the human genome can be applied equally effectively to the genomes of livestock. The ambition of the biomedical community is to sequence the entire human genome, whereas animal geneticists would be content with a detailed map of the livestock genomes.

The potential value of a map of the genome, together with the increasing ease with which very informative genetic markers can be identified, has led to the development of projects to build maps of domestic plant and animal species in order to further understand the genome and improve the tools for genetic selection. The Pig Gene Mapping Project is one such endeavor, and this chapter describes the rationale behind this project and its design and progress.

The pig

The pig (*Sus scrofa*) is one of the predominant meat-producing animals in Europe, North and South America, and Asia. For example, in Europe alone, pig production is worth approximately 17,500,000,000 European Currency Units (ECUs) annually. In many Asian countries, the pig is the major source of high-grade animal protein. The pig has also been subject to one of the more successful breeding programs within Europe. In the United Kingdom between 1970 and 1980, utilization of within-line selection and crossbreeding is estimated to have resulted in an improvement of more than 1% per annum in the proportion of lean in the carcass, with total annual benefits of around £100,000,000 (Mitchell et al. 1982).

From a genome analysis point of view, the pig has a number of advantages over other farmed mammalian species. The pig produces large litters of ten or more piglets and has an annual generation interval and a gestation period of 114 days, allowing more than two litters per year; thus experimental pedigrees can be produced rapidly. Like other farm mammals, the pig suffers marked inbreeding depression and few inbred lines are available; nevertheless, a number of very different outbred breeds are available within the species. In particular, European breeds such as the British Large White (Fig. 1, top) are genetically very different from Asian breeds such as the Chinese Meishan (Fig. 1, bottom). These differences are apparent both in a number of traits of economic and biological importance (Table 1) and at the genotypic level (Oishi et al. 1989). This means that crosses between pairs of the diverse breeds are

Table 1   European Large White and Chinese Meishan breed means and breed difference for commercially important traits

Trait	Meishan	Large White	Breed difference
Age at puberty (female, days)	115	180	4.0
Number of teats (female)	17	14	2.5
2nd parity litter size (female, piglets)	15	10	1.5
2nd parity ovulation rate (female, ova)	22	17	1.7
Prenatal survival (female, %)	85	65	1.0
Piglet birth weight (female, kg)	0.9	1.3	2.0
Growth rate (male, g/day)	381	724	6.0
Subcutaneous fat depth (female, mm)	26	9	4.5
Chop muscle area (female, cm^2)	17	37	4.0
Age at 80 kg (male, days)	220	150	6.0

The breed difference is given in within-breed phenotypic standard deviations. The sex in which the measure was recorded is indicated (Sources: Haley and Lee 1990; Haley et al. 1990a, 1992; Serra et al. 1992).

likely to be segregating for simple-sequence (i.e., RFLP) polymorphisms, as well as at VNTR loci such as minisatellites and microsatellites, in a manner similar to the interspecific crosses, which are so valuable for mapping in the mouse (Nadeau 1989). In addition to markers, QTL for most traits of economic importance will be segregating in crosses between the breeds, providing a resource for the study of these traits. Crosses between these diverse pig breeds are completely fertile (see Fig. 2).

The pig has a well-characterized karyotype of 18 pairs of autosomes plus the X/Y pair (Fig. 3). Of these chromosomes, 13 pairs are metacentric and 6 are acrocentric (see Committee for the Standardized Karyotype of the Domestic Pig 1988; Gustavsson 1990). (Some strains of wild boar have only 17 pairs of autosomes, with a centric fusion between chromosomes 15 and 17 [McFee et al. 1966; Popescu et al. 1989].) The karyotype is polydisperse, facilitating physical mapping techniques such as in situ hybridization and chromosome sorting. The total size of the porcine genome is not yet known with certainty, but it is thought to be similar to that of the human genome (i.e., 30 Morgans and 3×10^9 bp in length). The total DNA content of porcine leukocytes has been estimated to be 6.854 pg compared to 6.848 pg for human leukocytes (Mandel et al. 1950). No easy method of obtaining stable immortalized cell lines has yet been developed for the pig, but against this disadvantage must be weighed the ease with which large tissue samples can be obtained at slaughter, providing ample DNA for long-term studies.

Figure 1 (*Top*) European Large White sow and piglets. This breed and its derivatives are widely used worldwide. (*Bottom*) Chinese Meishan sow. This breed is currently restricted to the Lake Taihu region of China and some experimental herds in a few other countries. (Reprinted, with permission, from the Agricultural and Food Research Council.)

The pig gene mapping project

Recognition of its advantages has resulted in the pig's being the focus of the first multinational coordinated program for genome analysis in a mammalian farm species. The mapping of the porcine genome is being spearheaded by a multinational collaboration centered in Europe. Although most of the funding is being provided by national sources, the Pig Gene Mapping Project (or PiGMaP), funded to the extent of 1.2 million ECUs from the EC BRIDGE program, provides a focus for the research program and a framework for collaboration (Haley et al. 1990b; Archibald et al. 1991b; Haley and Archibald 1991). A wider grouping is provided by a European "laboratory without walls," which currently encompasses 17 laboratories in Belgium, Denmark, France, Germany, Italy, The Netherlands, Norway, Sweden, and the United Kingdom and may be further expanded in the future by the inclusion of research groups outside Europe.

The objectives of PiGMaP are to develop a 20-cM genetic (linkage) map covering 90% of the genome, to produce a physical map with at least one distal and one proximal landmark locus mapped on each chromosome arm, and to initiate the experiments necessary for mapping the QTL.

Figure 2 F_1 sow (Meishan x Large White) with an F_2 litter of piglets. (Reprinted, with permission, from the Agricultural and Food Research Council.)

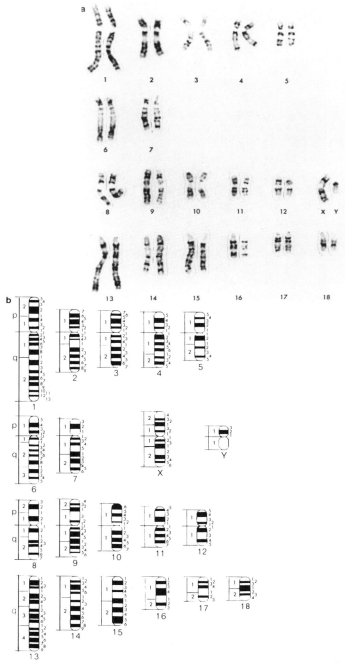

Figure 3 The porcine karyotype. (a) GTG-banded karyotype of a normal boar. (b) Schematic representation of the GTG-banded karyotype. (Reprinted, with permission, from the Committee of the Standardised Karyotype of the Domestic Pig 1988 and I. Gustavsson and *Hereditas*.)

GENETIC MAPPING STRATEGY

Genetic mapping in humans perforce relies on natural families. Information is maximized by selection of large full-sibships and their parents and preferably grandparents, and by the use of markers with high within-population heterozygosity. At the other extreme, some mapping in plants can utilize crosses between inbred lines of different species and use only RFLP markers that are fixed for different alleles in the two lines. The situation in pigs lies between these two extremes. Inbred lines from different species are not available, but genetically divergent breeds are available within the species. Furthermore, these breeds are of great interest to the quantitative geneticist and to the animal breeder, since they differ markedly for a number of traits (Table 1), and each breed contains genes of value to the other. In the first instance, map development and the study of QTL will rest upon crosses between divergent breeds. Most pig breeding, however, uses within-breed selection to improve performance, and thus markers that are informative within breeds will ultimately be required.

The strategy adopted for genetic mapping is to build a map based on crosses between divergent breeds and to incorporate markers into this map that are also informative within breed. The skeleton of the map will use RFLP markers based on homologous and heterologous sequences, many of which contain peptide-coding sequences. These markers usually have only two alleles and may be completely uninformative within a breed, but they can be highly informative in a cross between lines fixed for different alleles. The human or mouse homologous locus will already be mapped for many of these markers, and thus it will be possible immediately to start aligning the porcine map with maps of these other species. The skeleton of RFLP markers will be fleshed out with hypervariable VNTR-based markers, both those with minisatellite core sequences and those with microsatellite core sequences. The outline map based on polymorphisms detected with cDNA probes may also form the starting point for the development of a more detailed expression or transcriptional map of the porcine genome (see Hochgechwender 1992).

Reference families

The breeds of pig used for the reference crosses are of three types: European commercial breeds (Pietrain or Large White, see Fig. 1, top), Chinese Meishan (Fig. 1, bottom), and European wild boar. Five laboratories within PiGMaP will be contributing reference families from divergent crosses, each laboratory using one type of cross. Three laboratories, including our own, are working with Meishan x Large White crosses and two are working with wild boar x European commercial breed crosses. In the Edinburgh pedigrees, eight grandparental purebred pigs (two of each sex from each of the two breeds) were crossed reciprocally to produce the F_1 parental generation, members of which were intercrossed to yield F_2 individuals for the linkage mapping (see scheme in Fig. 4). In these three-generation families, the grandparents provide information on linkage phase, whereas the F_2 offspring provide information on the parental (F_1) meioses.

The reference families being produced by the five PiGMaP laboratories are thus F_2 intercrosses, although as the breeds are outbred, they only behave as classic F_2 populations for markers for which the two breeds are fixed for alternative alleles. The decision to use F_2 populations was made in part because for codominant markers fixed for alternative alleles in the two breeds, the F_2 provides more information on linkage than a backcross (Table 2). For markers with similar allele frequencies in the two breeds, the F_2 and the backcross (and indeed crosses within breed) provide similar amounts of information. Table 2 also shows that markers fixed for different alleles in the two breeds provide much more information than diallelic markers with heterozygosities of 0.5 (polymorphic information content [PIC] = 0.375) in each of the two breeds and also reaffirms the value of multiallelic markers.

Another reason to use F_2 intercrosses rather than backcrosses for mapping is the intent to map QTL in the same crosses. Several of the traits of interest show marked heterosis in the F_1 cross between the breeds, indicative of directional dominance at individual QTL. To map these QTL individually, a backcross to the breed carrying the recessive

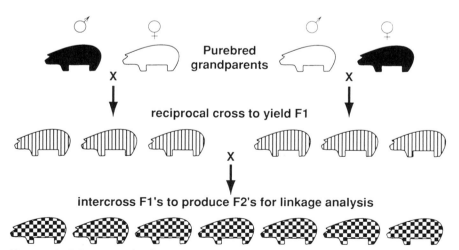

Figure 4 Schematic of a three-generation reference family: parental purebreds, F_1s, and F_2s. (Reprinted, with permission, from the Agricultural and Food Research Council.)

Table 2 Expected LOD scores per individual for linkage between pairs of different types of markers

Type of markers	Pedigree	Recombination fraction					
		0.0	0.05	0.1	0.2	0.3	0.4
Fixed for different alleles in two breeds	Backcross	0.301[a]	0.215	0.160	0.084	0.036	0.009
Fixed for different alleles in two breeds	F_2	0.452[a]	0.297	0.208	0.098	0.039	0.009
Two alleles segregating in two breeds[c]	F_2, backcross or within breed	0.038[b]	0.033	–	0.009	–	–
Five alleles segregating in two breeds[c]	F_2, backcross or within breed	0.211[b]	0.178	–	0.067	–	–
Ten alleles segregating in two breeds[c]	F_2, backcross or within breed	0.348[b]	0.278	–	0.134	–	–

[a]Theoretical values.
[b]From the analysis of simulated data, with full-sibships of size 8 plus parents and grandparents.
[c]Alleles at equal frequencies in both breeds.

allele would be a more powerful cross than the F_2, but for such loci, a backcross to the breed carrying the dominant allele contains no information at all. The direction of heterosis, and thus directional dominance, varies from trait to trait. For example, in a cross between the Meishan and the Large White, dominance is in the direction of Large White performance for growth rate (Bidanel et al. 1990; Haley et al. 1992), but in the direction of the Meishan for reproduction traits (Bidanel et al. 1989; Haley and Lee 1990). Furthermore, the direction of dominance may vary between loci controlling a single trait. Thus, the F_2 provides the best single cross for simultaneous study of several traits. Figure 2 shows the segregation of color and patterning seen in a litter of F_2 piglets from a cross between the Meishan and the Large White.

The initial mapping population will be made up of ten F_2 families, each of ten full-sibs and derived from a total of 20 grandparents. Two full-sib F_2 families are being provided by each of the five laboratories producing crosses. The genetic (linkage) map is being developed by a group of ten laboratories within the PiGMaP collaboration. The mapping strategy requires the typing laboratories to genotype the shared pool of animals. It is more efficient for each laboratory to type all reference family animals for each of the markers developed by that laboratory than for the groups holding animals to collect all of the primers and probes and run them across their own animals.

The reference population of 100 F_2 individuals is a powerful resource for detection of linkage for recombination fractions of at least 0.2 and will allow development of the initial map. Stored DNA samples from reference family animals will allow placement of new markers on the map as they are isolated. Markers will be selected from this map that will provide the reference markers for further mapping studies. These markers are likely to be largely based on microsatellite sequences because (1) microsatellite markers are likely to be polymorphic and thus useful in a variety of breed crosses and within breed, (2) further studies, particularly QTL mapping, will require genotyping large numbers of animals and the prospects for automated genotyping of microsatellite markers seem good, and (3) microsatellites provide sequence-tagged sites for physical mapping.

Selection of markers

Markers are selected which will have heterozygosity in the F_1 cross of ≥ 0.5 (equivalent to a PIC ≥ 0.375). In our own laboratory, markers are screened on four individuals per breed (two animals of each sex), which allows identification of those markers useful in a breed cross that are also likely to be informative within individual breeds. The animals screened are also the grandparents of the reference crosses, and thus markers selected for use in the cross must have within-breed heterozygosities of ≥0.5 (if the two samples have similar allele frequencies) in order to achieve a heterozygosity of ≥0.5 in the F_1. Alternatively, a heterozygosity of ≥0.5 in the F_1 can be achieved if the frequency of the most common allele in the two breeds differs, the required frequency difference being at most 0.5. With four individuals sampled per breed, the probability of detecting a within-breed polymorphism at a marker locus at which the population allele frequencies are equal is greater than 99% for loci with two alleles and increases further as the number of alleles increases. Thus, markers with a PIC ≥ 0.375 within one of the breeds are very unlikely to be missed in this screen, although within-breed heterozygosities for the selected markers will be poorly estimated and require reestimation in larger samples.

Reference markers will be selected at approximately 20-cM intervals. For pairs of the most informative markers, the standard deviation on an estimated recombination fraction of 0.2 will be about 0.03 with 100 F_2 animals (Mather 1951). Thus, there will be some variation in population recombination fractions, which are estimated to be 0.2 in the initial sample. This is of little concern, however, because (1) the power of interval mapping to detect QTL is not greatly affected by small variation in the distance between markers (Lander and Botstein 1989) and (2) the further data collected in the course of QTL mapping will allow a much more precise estimation of map distances for the selected reference loci. Where distances between adjacent reference loci turn out to unacceptably large, further markers found to map in approximately the right region in the reference families can be mapped more precisely in larger samples and selected if better positioned.

The precise mapping of markers enabled by the large samples required for mapping QTL will also allow study of differences in recombination fraction between the sexes and how this difference varies from region to region. Multi-allelic markers are ideal for this type of study as different heterozygous genotypes in a sire and dam allow the occurrence of recombinational events in the two sexes to be distinguished. In an F_2 cross, markers fixed for different alleles provide no information on sex differences in recombination frequency because both F_1 parents have the same genotypes, and thus it is not possible to distinguish whether a recombinational event occurred in the male or the female. In addition to study of sex differences, large sample sizes will allow the study of between-breed and even between-family variation in recombination frequency.

MARKERS: TYPE AND IDENTIFICATION

Three classes of molecular genetic markers are being employed in the genetic mapping studies. First, expressed sequences are being used in Southern blot analyses to detect diallelic RFLP loci (Archibald and Cowper 1990; Couperwhite et al. 1992a,b; Mariani et al. 1991a,b). Both homologous and heterologous (human, rodent, and other mammalian) probes (mainly cDNA) are being used to develop these RFLP markers. The Human Genome Mapping Project (HGMP) Resource Centre's DNA Probe Bank (Northwick Park, United Kingdom) has proved to be very useful for this endeavor. These expressed sequences will provide the means of integrating the genetic and physical maps as well as exploring comparative aspects of gene mapping. DNA prepared from the four animals of each breed that constitute the Edinburgh grandparents has been digested with a panel of 12 restriction endonucleases and is being screened with a range of homologous and heterologous probes. The diverse origins of the founding breeding stocks mean that many loci screened in this manner do indeed prove to be polymorphic.

The optimum RFLP marker locus is one for which the founder breeds (or grandparents) are fixed for different alleles. Several combinations of enzymes and probes have revealed such ideal polymorphisms. For example, among the porcine cDNAs that reveal such fixed differences between the Edinburgh Meishan and Large White grandparents are those which encode albumin (*ALB*), thyroid-stimulating hormone β subunit (*TSHB*), and protein phosphatase 2A β regulatory subunit (*PPP2ARB*) (Archibald et al. 1991a; Couperwhite et al. 1992a,b). The segregation of the *TSHB* alleles is illustrated in Figure 5. Other RFLP marker loci, including the casein loci and γ fibrinogen (*FGG*), are close to optimum, with all four grandparents of one breed fixed for one allele and three of the four founders of the other breed fixed for an alternative allele.

These diallelic RFLP loci detected with cDNA probes fit one of the criteria for type I anchor loci as defined by O'Brien (1991); i.e., they mark evolutionary conserved coding genes. Many of these loci also fit

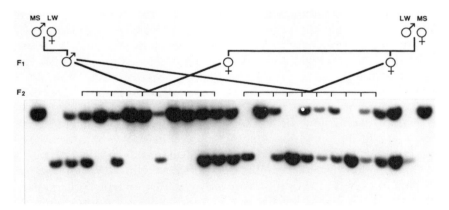

Figure 5 Segregation at the *TSHB* locus in the Edinburgh reference families. (Reprinted, with permission, from Couperwhite et al. 1992b.)

the other criteria of being available in the public domain and having been used to reveal homologous sequences in other species. The requirement for equivalence of the spacing of these loci in both the human and murine genomes (O'Brien 1991) will become of greater importance as more loci are mapped.

Although these RFLP loci will be invaluable for bringing the genetic (linkage) and physical maps together and for aligning the porcine gene map with the maps of other species, they have significant limitations for genetic mapping in the pig. For example, RFLP loci for which the founder breeds are fixed for different alleles will be useless for future within-breed mapping projects. Thus, hypervariable markers based on both types of VNTR loci—minisatellite and microsatellite loci—are also being developed. The presence of minisatellite loci in the pig has been demonstrated by the use of three different probes to reveal DNA fingerprint patterns (Georges et al. 1988). Coppieters et al. (1990) have described a porcine DNA fragment that behaves like an authentic minisatellite, revealing a DNA fingerprint at low stringency and a locus-specific pattern at higher stringency. The tandem repeats described within the glucose isomerase gene by Davies et al. (1992) are variable in number, but they fail to reveal a fingerprint at low stringency with pig genomic DNA. It is not yet known whether the minisatellite class of VNTR loci will be evenly distributed in the pig genome or whether they will be clustered as they are in the human genome (Royle et al. 1988) and in the bovine genome (Georges et al. 1990).

Most of the effort on VNTR marker development in PiGMaP is focused on the microsatellite loci. Wintero et al. (1992) have isolated and sequenced more than 100 loci containing simple tandem repeat $(dCdA)_n$ motifs. These authors have estimated that approximately 65,000–100,000 such loci exist in the porcine genome. By using the primed in situ labeling (Koch et al. 1989) of chromosomes with a $(dGdT)_n$ primer, they also demonstrated that these loci are evenly distributed throughout the genome, with only the regions around the telomeres, centromeres, and the nucleolar organizers showing any deficiency in such repeats. The interstitial heterochromatic bands at 16q21, 1q21, and Yq also appear to lack these repeat motifs. The porcine microsatellites therefore appear to have the same desirable characteristics as their human and murine equivalents. More than 200 porcine loci have now been sequenced, and several have been shown to be highly polymorphic both between and within breeds in several of the PiGMaP laboratories, including our own (Archibald et al. 1991b; Johansson et al. 1991; J.F. Brown and A.L. Archibald, unpubl.). The microsatellite loci will have a critical part to play in meeting our objective of 90% coverage of the porcine genome at the 20-cM level by the end of 1993. In the future, multiplex polymerase chain reaction (PCR) amplification of microsatellite loci and subsequent analysis of the products (or alleles) with automated DNA sequencers will be essential for the execution of QTL-mapping experiments involving several hundred to a few thousand animals. The VNTR loci, whether they are based on mini- or microsatellite sequences, fulfill O'Brien's (1991) criteria for type II anchor loci, i.e., they are high-resolution polymorphic genetic markers.

PHYSICAL MAPPING

The number of porcine chromosomes was correctly determined in 1931, but it is only in the past decade with the use of banding technology that an internationally agreed standard karyotype has been developed (Committee for the Standardized Karyotype of the Domestic Pig 1988) (see Fig. 3). Three physical mapping methods have been used for the assignment of genes to chromosomes in the pig: in situ hybridization, analysis of somatic cell hybrid lines, and analysis of flow-sorted chromosomes. The first porcine locus mapped by in situ hybridization was an *SLA* (swine lymphocyte antigen) class I gene that was assigned to chromosome 7 (Rabin et al. 1985; Echard et al. 1986). This first assignment indicated that heterologous probes can be used not only for Southern blot analysis, but also for in situ hybridization. The group at Institut National de la Recherche Agronomique, Toulouse, has also mapped the *TGFB, ENO1, APOE,* and *NP* loci using human probes (Yerle et al. 1990a,b). These studies confirmed the conservation of synteny of the *TGFB* and *APOE* loci with *GPI* and *RYR1,* and the conservation of the *ENO1* and *PGD* synteny. Thus, not only can information from the more detailed human and murine maps be used as a predictor for the location of porcine genes, but also cloned human (and murine) sequences can be used to confirm the predictions.

Although the polydisperse nature of the porcine karyotype facilitates chromosomal identification in somatic cell hybrids, this approach has not yet been as productive as it has been in human gene mapping. A full mapping panel of such cell lines has not been developed. Despite this limitation, a number of loci have been mapped using this methodology. In addition to X-linked genes such as *PGK, G6PD, GLA,* and *HPRT,* autosomal assignments have also been made (Leong et al. 1983; Ryttman et al. 1986).

The polydisperse nature of the porcine karyotype allows the chromosomes to be sorted effectively by the use of bivariate flow cytometry. The chromosomes are tagged with two different fluorochromes: Hoechst 33258, which is specific for A-T base pairs, and the G-C-specific chromomycin A3. The flow sorting of pig chromosomes has been achieved by groups in France and Cambridge (Grunwald et al. 1986; Dixon et al. 1992). Two lasers are used to excite these two fluorochromes, and thus the pig chromosomes can be resolved into 19 peaks in female samples and 20 peaks in male samples. Bouvet et al. (1992) have begun the determination of the chromosomal identity of the flow-sorted peaks, which should be completed within the next few months. The flow-sorted material has already been used to develop a chromosome-1-specific library, and other such libraries are planned (Miller et al. 1991, 1992). The characterization of porcine short interspersed (or SINE) sequences (Frengen et al. 1991) and the development of SINE-PCR techniques equivalent to *Alu*-PCR have facilitated the creation of these libraries from limited starting material.

It should also be possible to use the flow-sorted chromosomes for locus assignments. The sorted material can be addressed either by probing dot blots or by PCR amplification. The former approach requires larger numbers of chromosomes and therefore prolonged sorting times and is susceptible to the vagaries of dot-blot hybridizations. Although the PCR approach requires less starting material, it is exquisitely sensitive to contamination of the chromosome preparations, and although the image-enhanced presentation of a bivariate flow karyotype can show 19 or 20 discrete peaks or clusters, these peaks may be contaminated with fragmented chromosomes. Nevertheless, Chardon et al. (1991) have successfully assigned the *CYP21* and *TNFA* loci to chromosome 7 by PCR amplification of small numbers of sorted chromosomes (300–1000). If such assignments can be effected routinely, then flow-sorted chromosomes could supplant the role of somatic cell hybrids. The burden of maintaining and constantly validating the karyotypes of these inherently unstable cell lines would be avoided. For more precise localizations, in situ hybridization is currently the method of choice.

CURRENT STATUS OF THE PIG GENE MAP

The content of the pig gene map is constantly changing, and we do not seek to document its current status here. Rather, we have sought to identify the rationale for mapping the porcine genome and to highlight the distinctive aspects of PiGMaP. Geneviève Echard (Toulouse) maintains summaries of the pig gene map that are published regularly in *Genetic Maps* (Echard 1990), and Fries et al. (1990) have recently reviewed the state of the porcine map. The databases that are being developed in the context of PiGMaP should provide an accessible and up-to-date picture of the pig gene map as it is assembled. For example, in collaboration with staff at the Jackson Laboratory, a summary database, "PiGBASE," is being established in Edinburgh. PiGBASE can be seen as a porcine version of GBASE, the Jackson Laboratory's mouse genome database, or as an annotated bibliography concerned with pig gene mapping. Published linkage data and physical mapping data will be held in the database along with details of the relevant publications. At HGM10, the number of mapped porcine genes was 40, having changed little from the 38 recorded at HGM9 (Lalley et al. 1987, 1989). By HGM11, the number of mapped genes had risen to 80 (O'Brien and Graves 1991) and by the second PiGMaP meeting held in Toulouse 3–4 months later, the total was approaching 100. Of the 18 autosomes, only 3 (11, 17, and 18) currently lack assignments.

The linkage map of the pig, as summarized by Echard (1990), comprises seven linkage groups. The two best-characterized linkage groups, centered on the *HAL* and *SLA* loci, have been assigned to chromosomes 6 and 7, respectively, through in situ hybridization localizations of one or more members of the group. Similarly, linkage group V, encompassing the *TF*, *CP*, and *ELF* loci, has recently been mapped to chromosome 13. The predictable, but nevertheless poorly documented, close linkage of the casein loci (LG-IV) has recently been confirmed (A.L. Archibald et al., unpubl.). The genetic mapping consortium within PiGMaP is likely to effect a rapid expansion of the linkage map of the pig within the next 2 years. A pilot experiment involving seven laboratories and 20 F_2 pigs was carried out in the 6 weeks prior to the second PiGMaP project meeting held in Toulouse in December 1991. A total of 35 loci were genotyped and 4 new linkage groups were identified. For example, the following loci were assembled into a linkage group: the casein loci, the fibrinogen loci, one of the loci responsible for coat color, and the albumin locus (A.L. Archibald et al., unpubl.). Through linkage to the albumin locus, this group can be assigned to pig chromosome 8 (B.P. Chowdhary et al., in prep.).

INTEGRATION OF HIGH-RESOLUTION PHYSICAL AND GENETIC MAPS

For future QTL-mapping studies, it would be desirable to develop fully integrated physical and genetic (or linkage) maps of the porcine genome. It would also be desirable to align the integrated gene map with those of other species, especially the human and mouse.

Large-fragment genomic libraries will be useful in the development of integrated high-resolution physical and genetic maps. The necessary large-fragment libraries are being created in yeast artificial chromosomes (YACs) in London and Oslo and in P1 phage in Edinburgh. Because our ambitions are limited to mapping the porcine genome rather than to sequencing, our primary interest is in the 50,000–100,000 expressed genes. Expressed sequences (cDNA clones either "cloned by phone" or selected at random from gridded libraries derived from brain or total pig tissue) will be partially characterized by sequencing. These partially characterized cDNA clones will be used to screen the gridded large-fragment libraries. The large genomic clones identified in this manner will be used for physical (cytogenetic) and genetic (linkage) mapping. The entire genomic clone or subclones thereof will be used for chromosomal localization by fluorescent in situ hybridization. Informative genetic markers will be developed from such genomic clones by searching for microsatellite loci either by direct sequencing or by subcloning. As this strategy is pursued, a series of large genomic fragments will be identified and partially characterized. The characterization will include an expressed tagged site (cDNA sequence or active gene) and a sequence-tagged polymorphic site. In time, multiple genes will be assigned to individual YACs and, to a lesser extent, P1 clones. This gridded resource will be invaluable once QTL are mapped.

QTL MAPPING

The development of a 20-cM map will provide the tools to address the more important biological questions. Most genes of agricultural importance (and many controlling susceptibility to human diseases) do not have a large enough effect on their own to produce qualitative differences between individuals; rather, variation in several or many genes combines to produce continuous or quantitative variation between animals. Tracking the inheritance of markers in populations whose performance is recorded should allow the determination of the architecture of the genetic control of production traits. The Meishan/Large White populations are a particularly valuable resource in this respect, as the Meishan differs from the Large White for many traits, having, for example, genes for higher litter size, lower growth rate, higher fatness, and greater docility (Table 1).

The identification of genes controlling quantitative traits will be valuable for a number of reasons. First, it will provide information about the genetic architecture of quantitative genetic variation with respect to the distribution of sizes of gene effect and their individual action and position in the genome and allow study of interactions between QTL and between QTL and the genetic background and environment. This information will allow the building of models that should give improved prediction of long-term selection responses in breeding programs. Second, mapping of QTL opens the way to their manipulation by marker-assisted selection, using genotypes for the flanking markers to select individuals carrying the desired alleles. For individual loci, this direct genomic selection will be much more efficient than selection via the phenotype, although further research is required to identify optimum ways of combining marker-assisted and phenotypic selection. Finally, mapping offers a means of integrating physiological and molecular studies of variation. At present, it is difficult to associate physiological variation causally with phenotypic variation, but the coincident mapping of genes controlling both phenotypic and physiological variation provides evidence of causality and allows the pleiotropic effects of loci to be studied. This may illuminate causes of variation not only in the pig, but also in other mammals, including humans, because it is not unlikely that genes controlling, for example, fatness in the pig have homologs involved in the etiology of obesity in humans. The light these studies shed on the physiological basis of phenotypic variation will also enhance prospects for the ultimate cloning of QTL, which in turn will facilitate their more detailed study and possible ultimate exploitation.

Crosses between divergent genetic lines provide the most powerful resource for the detection of QTL and thus for the study of gene action underlying quantitative variation. One reason for using a cross is that the history of many animal breeds is long-term selection combined with some inbreeding. This will tend to move genes with larger effects to extreme allele frequencies and may lead to their fixation; i.e., within a breed, variation due to the very genes that are easiest to detect by mapping will have been greatly reduced. For example, Smith (1982) calculated that in a typical pig breeding program, the allele frequency at a locus controlling 3% of the variation in the trait under selection would be changed from 0.2 to 0.8 in about ten generations (which could be a period as short as 10 years). The same change would only take five generations if the locus controlled 11% of the variation. Thus, there is only a small chance of detecting QTL of appreciable effect within breeds that have undergone long-term selection. On the other hand, in a cross between lines selected in different directions, the QTL with the highest heterozygosities are likely to be those with the largest effect that are at or nearing selection for alternative alleles in the individual lines.

A second reason for using a cross is that drift and founder effects will have caused allele frequencies at marker loci to be very different in the divergent lines, and thus markers are much more informative than in outbreeding populations. In the best case, 100% of F_1 individuals may be heterozygous for a QTL and two flanking markers, whereas the equivalent figure for an outbreeding population may be only 12.5% (assuming heterozygosities of 0.5 at the three loci). The problem of using outbred populations is further exacerbated by the need to infer linkage phase between the markers and a putative QTL, which is unnecessary in a cross if it is assumed that the QTL is fixed for different alleles in the lines crossed.

The efficacy of using "interval mapping" to locate QTL in line crosses has been demonstrated both theoretically and by experiment (see, e.g., Paterson et al. 1988; Lander and Botstein 1989; Hilbert et al. 1991; Jacob et al. 1991). Interval mapping involves considering pairs of adjacent markers in turn and maximizing the likelihood for a QTL at each position between them. The curve showing the log likelihood ratio at each point along the chromosome provides a graphical representation of the evidence for the presence of a QTL in each position (Lander and Botstein 1989).

Interval mapping has several advantages over the use of single markers for QTL detection. In line crosses, interval mapping provides some extra power for the detection of QTL when markers are not close together and when the QTL effect is large (Lander and Botstein 1989; S.A. Knott and C.S. Haley, in prep.); however, the advantage in power of using two or more markers should be greater in outbreeding populations. Interval mapping provides much more precise estimates of the position and effect of QTL than does use of single markers, and it is also relatively insensitive to failure of the assumption that the trait is normally distributed (S.A. Knott and C.S. Haley, in prep.).

For a QTL contributing a proportion p of the variance in a cross between two lines, the minimum expected LOD (ELOD) from interval mapping (obtained when the QTL is equidistant from two markers) in a sample size n is approximately (Lander and Botstein 1989; S.A. Knott and C.S. Haley, in prep.)

$$\text{ELOD} = \left(\frac{1-2\theta}{1-\theta}\right)\frac{n}{2} \log_{10} \left(\frac{1}{1-p}\right)$$

when the recombination fraction between markers is θ. Thus, a QTL responsible for 3% of the variance in a population of 600 F_2 animals would be expected to produce a mean LOD of more than 3, with markers spaced at most 20 cM (i.e., $\theta = 0.165$) apart. Such a QTL would explain about 0.5 of a standard deviation between two lines fixed for alternative alleles. It thus is possible that several such QTL could be responsible for the breed differences in the traits shown in Table 1 and could be detectable in a cross between the breeds.

Despite its advantages in maximizing the likelihood, interval mapping suffers from being computationally relatively demanding, and thus it is tedious to estimate parameters for several QTL simultaneously. This is unfortunate because, when mapping a single QTL, estimates of its effect and position are biased if there is a second QTL on the same chromosome and the two QTL are in linkage disequilibrium (as is likely in a line cross) (S.A. Knott and C.S. Haley, in prep.). This problem can be overcome by use of an interval mapping method based on least squares. The results of this method are virtually identical to those of the maximum likelihood method, but interval mapping is much more easily applied and can thus be used to search for two QTL simultaneously (Haley and Knott 1992). Figure 6 shows an example of the method applied to simulated data in which two linked QTL were present and located simultaneously. The method provides accurate estimates of effect and position for pairs of linked QTL, and it should facilitate the analysis of data from line crosses such as those presently being produced in pigs.

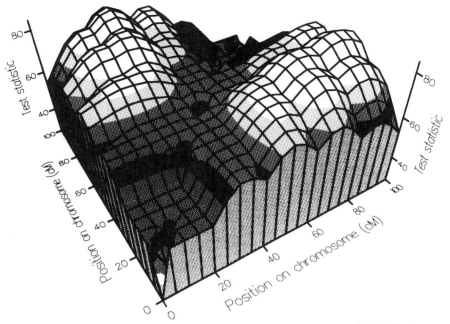

Figure 6 Simultaneous mapping of two QTL using regression methods. The horizontal axes represent the putative positions of the two QTL on a 100-cM chromosome, and the height of the surface at any point represents 2 $\log_e$ of the likelihood of two QTL in those positions. Two QTL positioned at 25 cM and 75 cM along the chromosome were simulated, and the analysis puts their most likely positions (the highest point in the surface) at approximately 30 cM and 90 cM along the chromosome. (Reprinted, with permission, from the Agricultural and Food Research Council.)

HOW CLOSE TO QTL CAN WE GET?

Two approaches to the cloning of mapped genes are the candidate gene approach and positional cloning (Wicking and Williamson 1991). Positional cloning currently involves immense application of resources and could not be contemplated in the pig. The candidate gene approach is preferable where it is possible and involves the exploration of genes that have already been cloned and are known to map to the same region as the target gene. This approach will become feasible in pigs as the porcine genetic map is developed and is aligned with the human map and as the identification of coding sequences proceeds in the Human Genome Project. Conservation of synteny between the human and pig sequences will make it possible to identify the region in the human sequence homologous to a region containing a QTL of interest in the pig. Most or all genes in that region in the pig are likely to have a human homolog in the same region, and thus the library of identified human sequences can be screened to identify likely candidate loci for the porcine QTL.

This approach sounds attractive, but a stumbling block will be the accuracy with which QTL can be mapped. Our simulation studies show that even with a 10-cM map of markers and using 1000 F_2 animals, a QTL responsible for 11% of the variance (i.e., contributing one standard deviation to the difference between two breeds fixed for different alleles) can only be mapped with an accuracy of about ±4 cM (S.A. Knott and C.S. Haley, in prep.). Assuming a map length of 30 Morgans with 100,000 randomly placed genes, this 8-cM section may contain approximately 300 loci. Even allowing that many of these loci can be discarded on the basis of their known or deduced function, screening the remainder would be an arduous task. Thus, much more efficient tools are required before cloning of QTL can be contemplated.

THE "HALOTHANE GENE": A PARADIGM FOR MAPPING MAJOR GENES IN LIVESTOCK

One of the most closely studied regions of the porcine genome is that around the HAL locus (for reviews, see Harrison 1979; Mitchell and Heffron 1980; Webb et al. 1982; Archibald 1991). The study of the molecular genetics of this locus offers for the future several important pointers of fine-scale gene mapping in livestock. For almost 20 years, the HAL locus has been known to have a major influence on important phenotypes. The HAL locus controls susceptibility to the so-called porcine stress syndrome, which has three manifestations: sudden stress deaths, pale soft exudative meat (PSE), and sensitivity to halothane. The sensitivity to halothane is manifested as a triggering of a malignant hyperthermic reaction on exposure to this anesthetic gas. Halothane sensitivity is taken to be a predictor of susceptibility to sudden stress death, since the deaths arise from malignant hyperthermic reactions to a variety of triggers, including mating, handling, and ambient temperature. Sensitivity to halothane-induced malignant hyperthermia has been shown to be controlled by a recessive gene at a single autosomal locus (*HAL*), with both alleles exhibiting variable penetrance. The halothane sensitivity allele (*HAL n*) is associated with beneficial effects on lean content and killing-out percentage, as well as deleterious effects on meat quality (e.g., PSE), litter size, litter weight, survival, and other traits (Webb and Simpson 1986; Simpson and Webb 1989). The beneficial effects explain the retention of this disease gene in commercial populations. The combination of the conflicting selection pressures and the inability of the halothane challenge test to discriminate between homozygote-resistant animals and carrier heterozygotes has also contributed to the maintenance of the disease allele.

Fortuitously, a number of biochemical polymorphisms are encoded by genes linked to the HAL locus. Thus, the HAL linkage group became one of the most intensely studied in the pig (see Fig. 7) (see Archibald and Imlah 1985; Archibald 1991). These biochemical genetic markers have been employed in one of the best examples of marker-assisted selection in livestock improvement to effect a reduction in the *HAL n* allele (Gahne and Juneja 1985). These markers, however, do not provide direct access to the HAL gene itself.

Only a molecular genetic approach would allow the information from linked markers to be used as a starting point for isolating the disease gene (Archibald 1987). Through the cloning of the *GPI* and *PGD* genes, a number of linked molecular genetic markers were developed for the prediction of HAL genotypes by linkage/haplotype analysis, and the HAL locus was mapped to the long arm of chromosome 6 (Davies et al. 1987, 1988; Chowdhary et al. 1989; Brenig et al. 1990a,b; Yerle et al. 1990a,b; Archibald and Bowden 1991). These DNA-based markers could have served as a starting point for identifying the HAL gene by positional cloning. However, at this stage, a candidate gene was suggested.

MacLennan et al. (1990) proposed that the gene responsible for susceptibility to halothane-induced malignant hyperthermia in pigs (and humans) encodes the skeletal muscle sarcoplasmic reticulum Ca^{++} release channel protein also known as the ryanodine receptor (RYR1). Subsequently, MacLennan and his colleagues demonstrated that a C to T

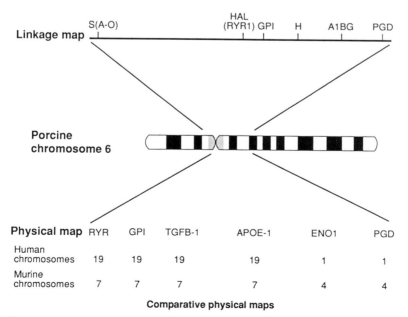

Figure 7 Genetic map of the halothane region. (Reprinted, with permission, from the Agricultural and Food Research Council.)

transition at nucleotide 1843 in this gene, leading to a substitution in the release channel protein of Arg-615 with cysteine, was correlated with susceptibility to malignant hyperthermia in five pig breeds (Fujii et al. 1991).

As a test of the candidacy of this mutation, one of us (A.L.A.) undertook a collaborative study with MacLennan and his colleagues to follow the segregation of the mutation in specific backcross families that had been created at our institute (Otsu et al. 1991). The pigs in these families had been subjected to a halothane challenge test and had been genotyped for some of the linked marker loci. In the backcross matings, all heterozygous pigs had the genotype *HAL N/n, GPI A/B, PGD A/B*, and all homozygous pigs had the genotype *HAL n/n, GPI X/X, PGD Y/Y*, where *X* and *Y* could be either *A* or *B*. Examination of the marker haplotypes indicated that all of the 11 (out of 376) disagreements between the RYR1 genotype (at nucleotide 1843) and the halothane test alone could be ascribed to animals that, although *HAL n/n*, did not react in the halothane test. Thus, the structure of the backcross families and the information provided by the linked genetic markers were invaluable in resolving the apparent discrepancies between the RYR1 genotypes and the halothane test results. In this study, 338 out of 338 informative meioses revealed cosegregation of the *HAL n* allele, with the C1843 to T mutation in the porcine RYR1 gene giving a LOD score for linkage at θ = 0.0 of 101.75.

Sarcoplasmic reticulum vesicles and single calcium release channels have been purified from pigs of differing malignant hyperthermia genotypes and studied in reconstituted lipid bilayer membranes. Calcium release channels from malignant hyperthermia susceptible pigs are slower to close after a contractile event, thus raising cytoplasmic calcium (Mickelson et al. 1988; Fill et al. 1990). Taken together, these data provide compelling support for the hypothesis that the C1843 to T mutation is causal for malignant hyperthermia in pigs.

It remains to be determined whether this mutation is also responsible for the differences in fat levels observed in pigs of different HAL genotypes. This question might be addressed by a "reverse genetics" approach of reproducing the mutation by homologous recombination in embryo-derived stem cells, which would be used to generate appropriate transgenic animals, initially mice but later pigs. Alternatively, the HAL region of chromosome 6 might be scanned for QTL contributing to leanness by following the segregation of markers in this region in populations lacking the RYR1 C1843 to T mutation. The Edinburgh PiGMaP pedigrees offer such an opportunity, especially since the body fat of the founder breeds (the Meishan and Large White) differs markedly.

Some lessons can be learned from the "halothane" story. As in human genetics, a candidate gene approach is often more productive than positional cloning. Comparative mapping studies and synteny conservation will greatly facilitate the candidate gene approach. For example, one can readily imagine that a QTL is mapped to a 10- or 20-cM region in pigs (or cattle or sheep) and that by examination of the homologous regions of the more detailed human and murine maps a number of genes can be considered as candidates for the QTL. Informative linked genetic markers not only serve as a starting point for isolating the gene of interest, but also can be used in the short term to manipulate the gene of interest by marker-assisted selection.

The localization of the sites on chromosomes occupied by genes or identifiable sources of variation in measured traits offers candidates not only for transfer in transgenic manipulations, but also for use in marker-assisted selection, i.e., to select animals on their genotypes at marker loci as well as on their phenotypes, thereby making it possible to select only those portions of the genome that are required from, say, an unimproved type of animal.

With an increased understanding of the human and porcine genomes, one can envisage improving the efficacy of pigs employed as models in medical studies by judicious selection or manipulation of the genetic makeup of the pigs used.

OTHER MAJOR GENES

The halothane gene has been intensively studied because of its high frequency in some breeds and its major economic impact. Other major genes affecting many aspects of porcine development and health have been either identified or hypothesized, and much of the evidence has been reviewed by Ollivier and Sellier (1982); recently, Robinson (1991) has listed known or suspected genetic defects. There are, of course, a number of genes controlling breed characteristics, particularly color, with the white and agouti loci segregating in a cross between the wild boar and the Large White, and the dominant white and extension loci probably those responsible for the segregation of color in the cross between the Meishan and the Large White (see Fig. 2). Searle (1968) has hypothesized about the homology between porcine and other mammalian color loci, and further evidence will come when they are mapped. The genetic basis of a number of other defects has been studied to a greater or lesser extent, including melanotic skin tumors (Hook et al. 1979), osteochondrosis leading to leg weakness (see, e.g., Lundeheim 1987), congenital blindness (Koch et al. 1957), lymphosarcoma (McTaggart et al. 1979), and a heritable ventricular septal defect (Swindle et al. 1990). A number of studies have also been done on the genetic basis of resistance to communicable disease, including respiratory diseases (Lundeheim 1979), and resistance to certain strains of pathogenic *Escherichia coli*, known to be inherited as a simple recessive and due to the absence of a specific receptor protein (Gibbons et al. 1977).

Under normal commercial conditions, any seriously genetically defective animals would be culled from the herd, and there are few lines established with the explicit purpose of studying particular lesions. The size of the pig population (~ 1.5 x 10^7 pigs born annually in the United Kingdom alone), however, indicates that many new mutations occur annually. If these individuals were identified, e.g., through national or commercial pig breeding programs, informative pedigrees could rapidly be established for those lesions that were not lethal reproductively (a three-generation pedigree descended from the proband and with >100 grand-offspring would only take 2–3 years to establish), allowing the loci responsible to be placed on the porcine (and by synteny, on the human) map. Thus, the national pig population would provide a rich source of material that could shed light on developmental processes in mammals and genetic disease in humans.

CONCLUSIONS

A gene map of the pig would provide a powerful resource allowing the detailed study of QTL controlling traits of economic importance for the first time. Mapping QTL could provide a means for manipulating them via marker-assisted selection and would aid their eventual isolation. Although cloning QTL will not be easy, it will be facilitated by a plentiful supply of candidate loci mapped in the Human Genome Project and by the synteny that exists between mammalian genomes. Identification of QTL will improve our understanding of the biological basis of quantitative variation in pigs, humans, and other species. Comparative mapping will shed light on genome organization and the evolutionary relationships between mammalian genomes, and should also allow models of human disease to be studied in the pig.

Acknowledgments

Our work on gene mapping in the domestic pig is supported with funds from the Agricultural and Food Research Council; the Ministry of Agriculture, Fisheries, and Food; and the Commission of the European Communities (BRIDGE programme). We thank our collaborators within the Pig Gene Mapping Project for their stimulating and entertaining company. Special thanks are due to Sara Knott, who produced Figure 6.

References

Archibald, A.L. 1987. A molecular genetic approach to the porcine stress syndrome. In *Evaluation and control of meat quality in pigs* (ed. P.V. Tarrant et al.), p. 343. Martinus Nijhoff, Dordrecht.

————. 1991. Inherited halothane-induced malignant hyperthermia in pigs. In *Breeding for disease resistance in farm animals* (ed. J.B. Owen and R.F.E. Axford), p. 449. C.A.B. International, London.

Archibald, A.L. and L. Bowden. 1991. Molecular markers for halothane-induced malignant hyperthermia in pigs. *Anim. Genet.* (suppl.) **22**: 97.

Archibald, A.L. and A.E. Cowper. 1990. Two *Taq*I RFLPs at the porcine pancreatic phospholipase A2 locus (PPLA2). *Anim. Genet.* **21**: 93.

Archibald, A.L. and P. Imlah. 1985. The halothane sensitivity locus and its linkage relationships. *Anim. Bld. Grps. Biochem. Genet.* **16**:253.

Archibald, A.L., J.F. Brown, S. Couperwhite, and C.S. Haley. 1991a. Reference families and markers for genetic mapping in the domestic pig. Human Gene Mapping 11. *Cytogenet. Cell Genet.* **58**: 2123.

Archibald, A.L., C.S. Haley, L. Andersson, A.A. Bosma, W. Davies, M. Fredholm, H. Geldermann, J. Gellin, M. Groenen, I. Gustavsson, L. Ollivier, E.M. Tucker, and A. Van de Weghe. 1991b. PiGMaP: An European initiative to map the porcine genome. *Anim. Genet.* (suppl.) **22**: 82.

Bidanel, J.P., J.C. Caritez, and C. Legault. 1989. Estimation of crossbreeding parameters between Large White and Meishan porcine breeds. I. Reproductive performance. *Genet. Sel. Evol.* **21**: 507.

————. 1990. Estimation of crossbreeding parameters between Large White and Meishan porcine breeds. II. Growth before weaning and growth of females during the growing and reproductive periods. *Genet. Sel. Evol.* **22**: 431.

Bouvet, A., N.G.A. Miller, P.D. Thomsen, and E.M. Tucker. 1992. Identification of chromosomes in pig flow karyotype. *Genet. Res.* (in press).

Brenig, B., S. Jürs, and G. Brem. 1990a. The porcine PHIcDNA linked to the halothane gene detects a *Hind*III and *Xba*I RFLP in normal and malignant hyperthermia susceptible pigs. *Nucleic Acids Res.* **18**: 388.

————. 1990b. The porcine PHIcDNA linked to the halothane gene detects a *Not*I RFLP in normal and malignant hyperthermia susceptible pigs. *Nucleic Acids Res.* **18**: 388.

Chardon, P., A. Schmitz, B. Chaput, G. Frélat, and M. Vaiman. 1991. SLA class III gene assignments combining sorting and polymerase chain reaction. *Anim. Genet.* (suppl.) **22**: 86.

Chowdhary, B.P., I. Harbitz, A. Aäkinen, W. Davies, and I. Gustavsson. 1989. Localization of the glucose phosphate isomerase gene to the p12→q21 segment of chromosome 6 in pig by *in situ* hybridization. *Hereditas* **111**: 73.

Committee for the Standardized Karyotype of the Domestic Pig. 1988. Standard karyotype of the domestic pig. *Hereditas* **109**: 151.

Coppieters, W., A. Van de Weghe, A. Depicker, Y. Bouquet, and A. Van Zeveren. 1990. A hypervariable pig DNA fragment. *Anim. Genet.* **21**: 29.

Couperwhite, S., B.A. Hemmings, and A.L. Archibald. 1992a. A *Bam*HI RFLP at the locus encoding the 65 kDa regulatory subunit of porcine protein phosphatase 2A (*PPP2AR*). *Anim. Genet.* (in press).

Couperwhite, S., Y. Kato, and A.L. Archibald. 1992b. A *Taq*I RFLP at the porcine thyroid stimulating hormone β-subunit locus (*TSHB*). *Anim. Genet.* (in press).

Davies, W., I. Harbitz, and J.G. Hauge. 1987. A partial cDNA clone for porcine glucosephosphate isomerase: Isolation, characterization, and use in detection of restriction fragment length polymorphisms. *Anim. Genet.* **18**: 233.

Davies, W., S. Kran, T. Kristensen, and I. Harbitz. 1992. Characterization of a porcine variable number tandem repeat sequence specific for the glucosephosphate isomerase locus. *Anim. Genet.* (in press).

Davies, W., I. Harbitz, R. Fries, G. Stranzinger, and J.G. Hauge. 1988. Porcine malignant hyperthermia carrier detection and chromosomal assignment using a linked probe. *Anim. Genet.* **19**: 203.

Dixon, S.C., N.G.A. Miller, N.P. Carter, and E.M. Tucker. 1992. Bivariate flow cytometry of farm animal chromosomes—A potential tool for gene mapping. *Anim. Genet.* (in press).

Echard, G. 1990. *Sus scrofa domestica* L. In *Genetic Maps* (ed. S.J. O'Brien), p. 4.110. Cold Spring Harbor Laboratory Press, Cold Spring Harbor, New York.

Echard, G., M. Yerle, J. Gellin, M. Dalens, and M. Gillois. 1986. Assignment of the major histocompatibility complex to the p1.4.q1.2 region of chromosome 7 in the pig (*Sus scrofa domestica* L.) by *in situ* hybridization. *Cytogenet. Cell Genet.* **41**: 126.

Fill, M., R. Coronado, J.R. Mickelson, J. Vilven, J. Ma, B.A. Jacobson, and C.F. Louis. 1990. Abnormal ryanodine receptor channels in malignant hyperthermia. *Biophys. J.* **50**: 471.

Frengen, E., P. Thomsen, T. Kristensen, S. Kran, R. Miller, and W. Davies. 1991. Porcine SINEs: Characterization and use in species specific amplification. *Genomics* **10**: 949.

Fries, R., P. Vögeli, and G. Stranzinger. 1990. Gene mapping in the pig. *Adv. Vet. Sci. Comp. Med.* **34**: 273.

Fuji, J., K. Otsu, F. Zorzato, S. de Leon, V.K. Khanna, J. Weiler, P.J. O'Brien, and D.H. MacLennan. 1991. Identification of a mutation in the porcine ryanodine receptor that is associated with malignant hyperthermia. *Science* **253**: 448.

Gahne, B. and R.K. Juneja. 1985. Prediction of the halothane (*Hal*) genotypes of pigs by deducing *Hal*, *Phi*, *Po2*, *Pgd* haplotypes of parents and offspring: Results from a large-scale practice in Swedish breeds. *Anim. Bld. Grps. Biochem. Genet.* **16**: 265.

Georges, M., M. Lathrop, P. Hilbert, A. Marcotte, A. Schwers, S. Swillens, G. Vassart, and R. Hanset. 1990. On the use of DNA fingerprints for linkage, studies in cattle. *Genomics* **6**: 461.

Georges, M., A.S. Lequarré, M. Castelli, R. Hanset, and G. Vassart. 1988. DNA fingerprinting in domestic animals using four different minisatellite probes. *Cytogenet. Cell. Genet.* **47**: 127.

Gibbons, R.A., R. Selwood, M. Burrows, and P.A. Hunter. 1977. Inheritance of resistance to neo-natal *E. coli* diarrhoea in the pig: Examination of the genetic system. *Theor. Appl. Genet.* **51**: 61.

Grunwald, D., C. Geffrotin, P. Chardon, G. Frélat, and M. Vaiman. 1986. Swine chromosomes: Flow sorting and spot blot hybridization. *Cytometry* **7**: 582.

Gustavsson, I. 1990. Chromosomes of the pig. *Adv. Vet. Sci. Comp. Med.* **34**: 73.

Haley, C.S. and A.L. Archibald. 1991. *The pig gene mapping project*. Commission of the European Communities DGXII, Brussels.

Haley, C.S. and S.A. Knott. 1992. A simple regression model for interval mapping in line crosses. *Heredity* (in press).

Haley, C.S. and G.J. Lee. 1990. Genetic components of litter size in Meishan and Large White pigs and their crosses. *Proc. 4th World Congr. Genet. Appl. Livest. Prod.* **15**: 458.

Haley, C.S., E. D'Agaro, and M. Ellis. 1992. Genetic components of growth and ultrasonic fat depth traits in Meishan and Large White pigs and their reciprocal crosses. *Anim. Prod.* **54**: 105.

Haley, C.S., G.J. Lee, I. Wilmut, C.J. Ashworth, R.P. Aitken, and W. Ritchie. 1990. British studies of the genetics of prolificacy in the Meishan pig. In *Proceedings of the Chinese Pig Symposium*, Toulouse, France (ed. M. Molénat and C. Legault). INRA, Jouy-en-Josas, France.

Haley, C.S., A.L. Archibald, L. Andersson, A.A. Bosma, W. Davies, M. Fredholm, H. Geldermann, M. Groenen, I. Gustavsson, L. Ollivier, E.M. Tucker, and A. Van de Weghe. 1990b. The pig gene mapping project: PiGMaP. *Proc. 4th World Congr. Genet. Appl. Livest. Prod.* **13**: 67.

Harrison, G.G. 1979. Porcine malignant hyperthermia. *Int. Anesthesiol. Clin.* **17**: 25.

Hilbert, P., K. Lindpainter, J.S. Beckmann, T. Serikawa, F. Soubrier, C. Dubay, P. Cartwright, B. De Gouyon, C. Julier, S. Takahasi, M. Vincent, D. Ganten, M. Georges, and G.M. Lathrop. 1991. Chromosomal mapping of genetic loci associated with blood-pressure regulation in hereditary hypertensive rats. *Nature* **353**: 521.

Hill, W.G. and S.A. Knott. 1990. Detection of genes of large effect. In *Advances in statistical methods for genetic improvement of livestock* (ed. K. Hammond and D. Gianola), p. 477. Springer-Verlag, Berlin.

Hochgeschwender, U. 1992. Toward a transcriptional map of the human genome. *Trends Genet.* **8**: 41.

Hook, R.R., Jr., M.D. Aultman, E.H. Adelstein, R.W. Oxenhandler, L.E. Millikan, and C.C. Middleton. 1979. Influence of selective breeding on the incidence of melanomas in Sinclair miniature swine. *Int. J. Cancer* **24**: 668.

Jacob, H.J., K. Lindpainter, S.E. Lincoln, K. Kusumi, R.K. Bunker, Y.-P. Mao, D. Ganten, V.J. Dzau, and E.S. Lander. 1991. Genetic mapping of a gene causing hypertension in the stroke-prone hypertensive rat. *Cell* **67**: 213.

Johansson, M., H. Ellegren, I. Edfors-Lilja, and L. Andersson. 1991. Cloning and characterization of highly polymorphic porcine microsatellites. *Anim. Genet.* (suppl.) **22**: 64.

Koch, P., H. Fischer, and H. Schumann. 1957. *Erbpathologie der Landwirtschaftlichen Haustiere.* Paul Parey, Berlin.

Koch, J.E., S. Kølvraa, K.B. Petersen, N. Gregersen, and L. Bolund. 1989. Oligonucleotide-priming methods for the chromosome-specific labeling of alpha satellite DNA *in situ. Chromosoma* **98**: 259.

Lalley, P.A., M.T. Davisson, J.A.M. Graves, S.J. O'Brien, T.H. Roderick, D.P. Doolittle, and A.L. Hillyard. 1987. Report of the committee on comparative mapping. *Cytogenet. Cell Genet.* **45**: 227.

Lalley, P.A., M.T. Davisson, J.A.M. Graves, S.J. O'Brien, J.E. Womack, T.H. Roderick, N. Creau-Goldberg, A.L. Hillyard, D.P. Doolittle, and J.A. Rogers. 1989. Report of the committee on comparative mapping. *Cytogenet. Cell Genet.* **51**: 503.

Lander, E.S. and D. Botstein. 1989. Mapping Mendelian factors underlying quantitative traits using RFLP linkage maps. *Genetics* **121**: 185.

Leong, M.M.L., C.C. Lin, and R.F. Ruth. 1983. The localization of genes for HPRT, G6PD and a-GAL onto the X-chromosome of domestic pig (*Sus scrofa domesticus*). *Can. J. Genet. Cytol.* **25**: 239.

Lundeheim, N. 1979. Genetic analysis of respiratory diseases of pigs. *Acta Agric. Scand.* **29**: 209.

———. 1987. Genetic analysis of osteochondrosis and leg weakness in the Swedish pig progeny testing scheme. *Acta Agric. Scand.* **37**: 159.

MacLennan, D.H., C. Duff, F. Zorzato, J. Fujii, M. Phillips, R.G. Korneluk, W. Frodis, B.A. Britt, and R.G. Worton. 1990. Ryanodine receptor gene is a candidate for predisposition to malignant hyperthermia. *Nature* **343**: 559.

Mandel, P., P. Métais, and S. Ciny. 1950. Les quantities d'acide désoxypentose-nucléique par leucocyte chez diverses espèceces de mammifères. *C.R. Acad. Sci.* **231**: 1172.

Mariani, P., M. Johansson, H. Ellegren, and L. Andersson. 1991a. A *Taq*I restriction fragment length polymorphism at the porcine platelet-derived growth factor β-receptor locus (PDGFRB). *Anim. Genet.* **22**: 446.

———. 1991b. *Taq*I and *Pvu*II restriction fragment length polymorphisms at the porcine plasminogen activator, urokinase, locus (PLAU). *Anim. Genet.* **22**: 447.

Mather, K. 1951. *The Estimation of linkage in heredity*, 2nd. edition. Meuthen, London.

McFee, A.F., H.W. Banner, and J.H. Rary. 1966. Variation in chromosome number among European wild pigs. *Cytogenetics* **5**: 75.

McTaggart, H.S., A.H. Laing, P. Imlah, K.W. Head, and S.E. Brownlie. 1979. The genetics of hereditary lymphosarcoma of pigs. *Vet. Rec.* **105**: 36.

Meyer, K. and W.G. Hill. 1991. Mixed model analyses of a selection experiment for food intake in mice. *Genet. Res.* **57**: 71.

Mickelson, J.R., E.M. Gallant, L.A. Litterer, K.M. Johnson, W.E. Remple, and C.F. Louis. 1988. Abnormal sarcoplasmic reticulum ryanodine receptor in malignant hyperthermia. *J. Biol. Chem.* **263**: 9310.

Miller, J.R., N.G.A. Miller, and E.M. Tucker. 1991. Construction of a chromosome 1-specific library from the domestic pig by universal amplification of FACS purified material. Human Gene Mapping 11 (1991). *Cytogenet. Cell Genet.* **58**: 2129.

Miller, J.R., S.C. Dixon, N.G.A. Miller, E.M. Tucker, J. Hindjkaer, and P.D. Thomsen. 1992. A chromosome 1-specific DNA library from the domestic pig (*Sus scrofa d.*). *Cytogenet. Cell Genet.* (in press).

Mitchell, G. and J.J.A. Heffron. 1982. Porcine stress syndrome. *Adv. Food Res.* **28**: 167.

Mitchell, G., C. Smith, M. Makower, and P.J.W.N. Bird. 1982. An economic appraisal of pig improvement in Great Britain. 1. Genetic and production aspects. *Anim. Prod.* **35**: 215.

Nadeau, J.H. 1989. Maps of linkage and synteny homologies between mouse and man. *Trends Genet.* **5**: 82.

O'Brien, S.J. 1991. Mammalian gene mapping: Lessons and prospects. *Curr. Opin. Genet. Dev.* **1**: 105.

O'Brien, S.J. and J.A.M. Graves. 1991. Report of the committee on comparative mapping. Human Gene Mapping 11 (1991). *Cytogenet. Cell Genet.* **58**: 1124.

Oishi, T., K. Tanaka, T. Otani, and S. Tamada. 1989. Genetic variations of blood groups and biochemical polymorphisms in Meishan pigs. *Bull. Natl. Inst. Anim. Ind.* **48**: 1.

Ollivier, L. and P. Sellier. 1982. Pig genetics. *Ann. Genet. Sel. Anim.* **14**: 481.

Otsu, K., V.K. Khanna, A.L. Archibald, and D.H. MacLennan. 1991. Cosegregation of porcine malignant hyperthermia and a probable causal mutation in the skeletal muscle ryanodine receptor gene in backcross families. *Genomics* **11**: 744.

Paterson, A.H., E.S. Lander, J.D. Hewitt, S. Peterson, S.E. Lincoln, and S.D. Tanksley. 1988. Resolution of quantitative traits into Mendelian factors by using a complete linkage map of restriction fragment length polymorphisms. *Nature* **335**: 721.

Popescu, C.P., J. Boscher, and G.L. Malynicz. 1989. Chromosome R-banding patterns and NOR homologies in the European wild pig and four breeds of domestic pig. *Ann. Genet.* **32**: 136.

Rabin, M., R. Fries, D. Singer, and F.H. Ruddle. 1985. Assignment of the porcine major histocompatibility complex to chromosome 7 by *in situ* hybridization. *Cytogenet. Cell Genet.* **39**: 206.

Robinson, R. 1991. Genetic defects in the pig. *J. Anim. Breed. Genet.* **108**: 61.

Royle, N.J., R.E. Clarkson, Z. Wong, and A.J. Jeffreys. 1988. Clustering of hypervariable minisatellites in the proterminal regions of human autosomes. *Genomics* **3**: 352.

Ryttman, H., P. Thebo, I. Gustavsson, B. Gahne, and R.K. Juneja. 1986. Further data on chromosomal assignments of pig enzyme loci LDHA, LDHB, MPI, PEPB and PGM1, using somatic cell hybrids. *Anim. Genet.* **17**: 323.

Searle, A.G. 1968. *Comparative genetics of coat colour in mammals.* Logos Press, London.

Serra, J.J., M. Ellis, and C.S. Haley. 1992. Genetic components of carcass and meat quality traits in Meishan and Large White pigs and their reciprocal crosses. *Anim. Prod.* **54**: 117.

Simpson, S.P. and A.J. Webb. 1989. Growth and carcass performance of British Landrace pigs heterozygous at the halothane locus. *Anim. Prod.* **49**: 503.

Smith, C. 1982. Estimates of genetic change in pig stocks with selection. *Z. Tierzuchtg. Zuchtsbiol.* **99**: 232.

Swindle, M.M., R.P. Thompson, B.A. Carabello, A.C. Smith, B.J.S. Hepburn, D.R. Bodison, W. Corn, A. Fazel, W.W.R. Biederman, F.G. Spinale, and P.C. Gillette. 1990. Heritable ventricular septal defect in Yucatan miniature swine. *Lab. Anim. Sci.* **40**: 155.

Webb, A.J. and S.P. Simpson. 1986. Performance of British Landrace pigs selected for high and low incidence of halothane sensitivity. 2. Growth and carcass traits. *Anim. Prod.* **43**: 493.

Webb, A.J., A.E. Carden, C. Smith, and P. Imlah. 1982. Porcine stress in pig breeding. *Proc. 2nd World Congr. Genet. Appl. Livest. Prod.* **6**: 588.

Wicking C. and R. Williamson. 1991. From linked marker to gene. *Trends Genet.* **7**: 288.

Winterφ, A.K., M. Fredholm, and P.D. Thomsen. 1992. Variable (dG-dT)n·(dC-dA)n sequences in the porcine genome. *Genomics* **12**: 281.

Yerle, M., A.L. Archibald, M. Dalens, and M. Gillois. 1990a. Localization of PGD and TGFβ-1 to pig chromosome 6q. *Anim. Genet.* **21**: 411.

Yerle, M., J. Gellin, M. Dalens, and O. Galman. 1990b. Localization on pig chromosome 6 of markers GPI, APOE, and ENO1 carried by human chromosomes 1 and 19, using *in situ* hybridization. *Cytogenet. Cell Genet.* **54**: 86.

YAC-based Mapping of Genome Structure, Function, and Evolution

David Schlessinger and Juha Kere

Department of Molecular Microbiology
Washington University School of Medicine
St. Louis, Missouri 63110

Human geneticists traditionally ask how to locate genes and understand their function, particularly genes that are modified in the causation of inherited diseases or cancer. Yeast artificial chromosomes (YACs) are helpful because they are large enough to recover large genes intact and to facilitate targeted coverage of megabase regions in overlapping clones. The process can be extended to whole chromosome coverage. Comparative maps can then be assembled across much of phylogeny with minimal sets of probes. As contiguity becomes more complete, global analytic approaches become increasingly attractive. These approaches include refined structural studies to the level of sequence, systematic screening for genes and candidate disease genes across regions, and inferences about the evolution of genes and genomes.

This chapter discusses:

❑ the way in which YACs facilitate the assembly of maps

❑ use of the resultant maps in the structural, functional, and evolutionary analyses of the human genome

INTRODUCTION

As the mapping of the human genome advances, analyses will turn increasingly toward the utilization of the information. The expectation is

Genome Analysis Volume 4: *Strategies for Physical Mapping*

that genomic map information will provide a basis for the understanding of genome function. In addition to the genes, the genome includes meaningful elements that function in various processes. These elements should include at least origins and control regions for replication, signals for chromosome assembly and pairing, hot spots for recombination, signals for X chromosome inactivation, and possible sites for DNA imprinting.

At present, most functional analyses are directed toward the identification of the control regions and transcription units of genes. However, genes in toto are expected to occupy only about 5–10% of the genomic sequence content (Bishop 1974; Bodmer 1981), and only about 2000 of the estimated 100,000 genes are known (Klinger 1991). At least 4000 of those genes are now thought to be involved in hereditary illnesses or cancer (McKusick 1990).

To focus on the impact of YACs and YAC contigs on mapping and its relationship to disease gene hunts, one can start from the four-step model described below and summarized in Table 1.

Step I: Identification of the region of interest

Genetics has become the indispensable first step that permits the localization, by linkage or cytogenetic studies, or both, of a gene or genes of interest to a delimited region of the genome. The region can be localized to a cytogenetic band (about 10 Mb) or even to about 1–2 Mb by genetic mapping (Donis-Keller et al. 1987); in favorable cases, the region of interest can be specified even more precisely by the detection of deletions or translocations, presumably disrupting a gene involved in pathogenesis. In fact, chromosomal abnormalities have been used to map most of the tumor suppressor genes and other cancer-related genes, as well as inherited disease genes found thus far. In rare cases such as cystic fibrosis, only genetic linkage information followed by physical mapping provided the route to the isolation of the gene (Kerem et al. 1989; Riordan et al. 1989; Rommens et al. 1989).

Step II: Cloning of the DNA in the region of interest

In some cases, "candidate genes" already mapped to the region can include the critical sought-after gene (McKusick 1991). Such genes, like translocations and deletions, provide shortcuts in map-based searches. More generally, starting from one or preferably more markers in the region (frequently the probes that detected polymorphism in genetic mapping efforts) and using a combination of walking and cloning procedures, the overlapping clones of DNA that span the region can be assembled.

Table 1 Current paradigm for positional cloning

Step I. Identification of region of interest
    Genetic linkage mapping
    Chromosomal abnormality (translocation, microdeletion, other
        rearrangement)
Step II. Recovery of cloned DNA from the region
    Nontargeted approaches:
        random probes
    Targeted approaches:
        microdissection (plasmids)
        irradiation hybrids, microclones (phages, cosmids)
        YAC libraries (YACs)
    Continuous coverage of targeted region (YAC contigs)
Step III. Search for putative coding sequences (Table 3)
    Sequence characteristics
    Expression studies
    Evolutionary conservation
    In vivo functional assays
Step IV. Identification of relevant gene(s)
    Genetic linkage (recombinants, allelic associations)
    Pathogenetic mutations vs. polymorphisms
    Relevant properties and suggested likely pathogenetic mechanism
  (expression pattern, alterations in gene product)
    Functional tests (transfections, transgenic animals)

The use of YACs impacts directly on Step II, where long-range continuity and isolation of additional probes are most efficiently achieved with YACs; methods are currently being developed to search YACs or YAC contigs directly for coding sequences (Step III); and the ability of YACs to harbor even large genes intact with their regulatory elements will increase their usefulness for functional tests in Step IV.

Step III: Search for putative coding sequences

Once the DNA is in hand, genes are searched for in the region of interest. A variety of different methods used for this purpose are summarized in Table 3 and discussed below. Generally, these methods are used in combination to ensure that the gene of interest is discovered.

Step IV: Identification of relevant genes

Genes are screened as candidates for the locus of a change that contributes to disease; every detected change is studied to discriminate indifferent polymorphism from pathogenetic mutation. At this step, genetic linkage once again becomes a valuable tool. Once the potential mutation has been detected, effects of the modified gene are open to study, often utilizing transgenic animals to model the pathophysiology in vivo.

    The development of YACs as a cloning vector (Burke et al. 1987) and the progress of the Human Genome Project (National Research Council 1988) have been important advances that directly affect Steps II and III; accumulation of the map information and resources will increasingly facilitate these steps. Methods for the use of YACs for Step IV are likely to develop further (D'Urso et al. 1990; Eliceiri et al. 1991; Gnirke et al. 1991; Huxley et al. 1991).

IMPACT OF YACS ON THE GENERATION OF LONG-RANGE CONTIGUOUS CLONED DNA

The power of YACs is most clear in projects that seek to find DNA across a region greater than 100 kb. Previous cloning systems using bacterial plasmid or phage vectors had two limitations. One was the limited size of clones; with the development of cosmid vectors the upper limit had arrived at about 40 kb and remained unchanged for about 10 years. The second limitation was the occurrence of unclonable regions at an average distance of 100–200 kb. Because of these two deficiencies, overlapping clones required a number of walking steps to achieve as much as 200 kb of coverage and rarely extended much further before encountering a gap. Attempts were made to jump across gaps or infer their size by pulsed-field gel electrophoresis (Barlow and Lehrach 1987); however, the work was slow and labor-intensive, and one was left without continuous cloned DNA.

YACs remedy both of these limitations. First, individual YACs can be up to 20-fold larger than the largest cosmid clones, so that a single YAC easily exceeds the size of a typical cosmid/λ/plasmid contig. Second, YACs have proven to clone more than 99% of human DNA in a number of regions (Schlessinger et al. 1991). Consequently, contigs extend to the megabase scale and are assembled with many fewer steps. YACs thus affect the achievement of long-range contiguity in at least three ways as described below.

YACs can provide even large genes intact and in proper context

The sheer size of YACs has tended to make simple the isolation of large structural genes. In the most obvious application, transfection studies of intact hypoxanthine phosphoribosyltransferase (HPRT) (Huxley et al. 1991) and glucose-6-phosphate dehydrogenase (G6PD) (D'Urso et al. 1990) have been possible. In addition, the assay of tissue-specific differentiation signals or long-distance regulatory mechanisms in regions of the genome can be much more incisive, since control elements for an individual gene may lie in a large intron or in relatively distant upstream sequences that are frequently absent from typical cosmid constructs.

Similarly, when attempts are made to assay for genes like tumor suppressors by transfecting mutant cells with cloned DNAs, the large amount of DNA in a YAC greatly increases the likelihood that both the gene and its regulatory apparatus will be included. YACs provide the starting material that can be progressively subcloned, using the functional assay to isolate the gene.

An additional advantage of YACs for such purposes is provided by the high activity of yeast homologous recombination. In the case of the cystic fibrosis transmembrane regulator (CFTR) gene, for example, where in a starting library no YAC contained the entire gene, it was possible to recombine overlapping YACs to achieve clones with the gene intact (Green and Olson 1990). Comparable constructions have been successfully carried out for the Bcl-2 (Silverman et al. 1990) and Duchenne's muscular dystrophy (DMD) (Den Dunnen et al. 1992) genes as well.

YACs can provide coverage of 1-2 Mb targeted regions

YACs now provide a mechanism to obtain overlapping DNAs across a chromosomal region. The process is obviously simpler when a region is delimited to an extent that it can be included in one or a few YACs. For example, with a collection that contains five genomic DNA equivalents in YACs with an average size of 250 kb, a single screening with a probe detects an average of five YACs covering about 750 kb. Subsequent screenings can then be done with probes derived from the ends of the YACs extending out the furthest; as a result, three to five screenings can cover on the order of 2 Mb in overlapping clones.

An increasing number of such contigs are now being reported (Table 2). Such approaches and contigs are especially useful in the targeted search for disease genes, where a probe or bracketing probes, obtained by cytogenetic studies like in situ hybridization or detection of a deletion, genetic studies of linkage, or a combination of inferential methods like pulsed-field gel electrophoresis to compare features of the DNA in normal and affected individuals, define the region of interest.

YACs can provide long-range contiguity moving toward a total map

The contribution of YACs becomes much more decisive as sustained long-range mapping takes hold under the human genome initiative. Once maps across chromosomes provide an overall structural framework, a particular gene could be found, including one involved in the etiology of a specific disease, by simply asking for the YACs in the neighborhood of the initial probe(s) of interest.

Targeted coverage of a region is to long-range contiguity of the genome much as the search for genes one-by-one is to systematic searches for genes across the map. When the entire map becomes available, targeted cloning will become superfluous, but until that time, targeted cloning is indispensable. Fortunately, the two types of mapping are commensurate: They are always complementary, and they can usually proceed comfortably at the same time, with targeted cloning forming part of

Table 2 Examples of large YAC contigs from various regions of the genome

Region and genes	Contig size (Mb)	YACs	References
1q32, MCP, CR1, CR2, DAF	0.8	9	Hourcade et al. (1991)
6p21.3, HLA	2.5	42	Bronson et al. (1991); Kozono et al. (1991); Geraghty et al. (1992)
7q31,CFTR	2.5	50	Green and Green (1991)
11p13,WAGR	1	5	Bonetta et al. (1990)
18q21, BCL2, PLANH2	2	16	Silverman et al. (1991)
21q, D21S13, D21S16	1.5	9	Butler et al. (1992)
Xp21.2, DMD	2.6	34	Coffey et al. (1992)
Xq26, HPRT, F9	8	94	Little et al. (1992)
Xq28, F8	1.6	25	Freije and Schlessinger (1992)
Xq28, IDS	1.2	20	Palmieri et al. (1992)
Xq28, FRAXA	1.6	6	Hirst et al. (1991)

an overall strategy that recovers YACs for many regions, merging into a continuous map.

A formulation of the mapping process that is often useful begins with Maynard Olson's notion of "vertical integration" in genome analysis: smooth passage from the level of chromosomes to the level of sequence, with the ideal of a set of steps that proceed as systematically as possible. In general, one can proceed upward from the level of sequence and downward from chromosomes through YACs. However (see, e.g., Schlessinger 1990), pathways from chromosomal DNA to the level of YACs or from YACs toward sequence depend on elective technology. For example, sets of radiation hybrids (Cox et al. 1990), sets of fragmented human chromosomes (Farr et al. 1991), hybrid cell panels, panels of DNAs derived from patients with localized deletions or translocations (Ballabio et al. 1989), and in situ hybridization (Lichter et al. 1990; Montanaro et al. 1991) all offer ways to compartmentalize DNA fragments in subchromosomal regions. All of these methods are improving and can be used to assemble YACs in successive regions of a chromosome. At low resolution (i.e., lower doses of radiation, smaller numbers of fragmented chromosomes, in situ hybridization with prometaphase chromosomes, hybrid panels with few members), these methods produce compartments on the order of a cytogenetic band, within which YACs may then be overlapped. As resolution increases, with higher doses of radiation, larger numbers of fragmented chromosomes, and in situ hybridization in interphase with differentially labeled YACs, these methods all become smoother and more continuous. Of course, the resolution required depends in part on the quality and size of the YACs being used: With YACs on the order of 1 Mb in size—a value approached in some current libraries (see, e.g., Anand et al. 1990)—compartments need reach only that level of resolution.

In fact, the achievement of larger clones in recent collections of YACs exemplifies the direct relation of changes in technology to the constant goal of coverage of ever-larger regions of the genome in overlapping YACs with the minimum number of library screenings. As another example, the development of systematic methods to clone the ends of YAC inserts (Pfeifer et al. 1989; Riley et al. 1990; Rosenthal and Jones 1990; Lagerstrom et al. 1991; J. Kere et al., in prep.) has simplified both the characterization of YACs and the techniques for moving from one clone to overlapping clones. At present, a thousand flowers bloom in the field of YAC technology, and progress is tightly dependent on innovation both in YAC library construction and in mapping methods.

There are currently two important deficiencies in YAC cloning methods. One severe limitation is incomplete representation. Although very nearly all human DNA can be cloned into YACs, some of it has not been recovered by screening of a number of current libraries, and other segments (like long tandemly repeated sequences; Neil et al. 1990) show considerable instability, throwing off deletions during the outgrowth of clones. Such problems occur about every 2 Mb on average and may be more severe toward telomeres (Freije and Schlessinger 1992). They provide an upper limit on the extent of continuous coverage, at least until a way is found to stabilize such DNA in wild-type or mutant yeast cells.

The other major failing of YAC cloning is the tendency to create chimeric clones in which two segments of DNA originating from noncontiguous regions of chromosomal DNA are joined together. Cocloning is a major nuisance and seems to be promoted, at least in some cases, by the constitutive high levels of homologous recombination in yeast (Green et al. 1991a). One can nevertheless make some positive observations:

1. Different libraries of YACs show varying levels of such cocloning events (see, e.g., Schlessinger et al. 1991). Thus, the phenomenon does not appear to be intrinsic to YAC cloning but depends in a poorly defined way on the amount and handling of extraneous or carrier DNA during library construction. It now seems likely that the factors critical in this process should be discernible and that future YAC libraries will equal or better the current best libraries, which seem to contain on the order of 5% cocloning (Anand et al. 1990).

2. As in the case of cosmid libraries, the finding of good clones in a region permits the detection of scrambled clones from that region with fair accuracy. Libraries with multiple genome coverage are therefore essential.

3. Despite cocloning, YAC-based maps have been assembled in nearly every instance where they have been attempted.

Limitations are thus generally manageable and may be eliminated. Methods for the assembly of YAC contigs must satisfy two needs: (1) the generation of an ordered set of overlapping YACs and (2) the formatting of the resultant map. Map assembly is always based on the common sequence content of overlapping clones, but this can be assessed and characterized in many ways. Two major tools of the map assembler are DNA-DNA hybridization and the polymerase chain reaction (PCR).

Using hybridization, we can utilize probes (which may be single-copy, moderately repetitive, or highly repetitive) to screen collections of YACs for overlaps, and then use additional methods to confirm matches. For example, oligonucleotide or naturally moderately repetitive probes can be used against matrix arrays of DNAs from a series of YACs (Lehrach et al. 1990). With unique probes, the finding of overlapping YACs is straightforward, whereas with repetitive sequences, the hybridizing YACs must be subclassified to define a group. This can usually be done by a "fingerprinting" strategy. For example, DNA from each clone can be digested with one or two restriction enzymes, and the sizes of the DNA fragments that contain the repetitive sequence can then be determined in Southern blot analyses (Zucchi and Schlessinger 1992; G. Porta et al., in prep.). Related YACs are then recognized by their content of fragments with the same mobility. In the limited case, where a succession of common oligonucleotide or repetitive sequence probes is used, a statistical criterion must be applied to determine the probability of overlap versus the random chance concordance of fragment sizes that hybridize to that probe.

Such hybridization-based methods can be used to assemble YACs across localized or more extensive regions of a genome. Variations of fingerprinting have been used very successfully in the generation of a physical map of *Caenorhabditis elegans* (Coulson et al. 1991) and in targeted walking, starting from YACs seeded in subregions, to assemble all of the human contigs done thus far, including a set as large as 8 Mb in Xq24-qter (Little et al. 1992).

An example: Mapping of Xq24-q28

Map construction in this extensive example proceeded with a number of techniques based on hybridization (for review, see Schlessinger et al. 1991). First, YACs were localized in compartments, either by recovering them with single-copy probes already placed in subregions of Xq24-q28 or by assigning them to regions of cytogenetic bands (± 0.5) by nonradioactive in situ hybridization (Montanaro et al. 1991). Contig assembly then continued using a variety of additional probes to detect overlapping YACs, usually by hybridization to the collection arrayed in a gridded matrix on nylon membranes. The probes used included (1) end-clones or other probes made from YACs, which permitted "walks" to neighboring clones; (2) a YAC itself, which cross-hybridized to overlapping clones; and (3) repetitive sequences as fingerprinting probes.

All of the techniques proved serviceable, at least to some extent, because the YAC library was largely representative of the Xq24-q28 region. From the earliest stages of the analysis, the recovery of Xq sequences has been consistent with cloning that is esentially random, i.e., nearly all sequences are equally clonable in YACs.

The capacity to assemble contigs of up to 8 Mb (Little et al. 1992) itself demonstrates the extensive representation of genomic sequences in YACs. However, several more formal tests of representation have been applied. For example, several types of probes were tested by hybridization against the library of clones to determine how many cognate clones were recovered. The probes included some of those previously identified by the community of X chromosome workers: end-clones derived by subcloning, by *Alu*-vector PCR (Nelson et al. 1989), or by primer-ligation-mediated PCR (Riley et al. 1990); inter-*Alu* PCR products (Nelson et al. 1989) from YAC inserts; and a moderately repetitive sequence, pTR5, which occurs 50 times in the Xq24-q28 region (Zucchi and Schlessinger 1992).

Well over 90% of probes found at least one YAC in the collection. A further critical test of representation was carried out, based on additional evaluation of the null class. Were probes and end-clones for which no YACs had been recovered "unclonable" in YACs, or did they represent sequences absent from the library for statistical reasons? To test for the clonability of sequences missing from this YAC collection, a second library of YAC clones was screened. Probes were sequenced, and on the basis of the sequences, PCR primers were developed to screen the library of total genomic DNA in the Center for Genetics in Medicine. If cloning were indeed random, that library, which also contains three genomic equivalents of X DNA, would yield YACs for 90% of the probes not found in the first collection. This was in fact the result obtained.

This experience is consistent with several other studies that have achieved coverage of up to 2 Mb in other regions of the genome (Table 2) and is also in accord with results of screening the Center for Genetics in Medicine library with a series of 300 primer pairs for autosomal genes: More than 96% of those screenings have been positive. There are sequences that are either not cloned into YACs in a number of libraries or are unstable in YACs, but those regions are clearly limited in extent.

The current contigs over the Xq26-q28 region cover nearly 30 Mb of DNA in contigs, which are at least 1 Mb each and which are still growing. The Xq24-q28 work has demonstrated that YAC-based mapping can be essentially as complete as desired. Representation is nearly total, with very nearly all probes finding YACs. Furthermore, closure can be effected in most cases simply by screening additional collections of YACs with probes that have found no YACs in a given library. In the regions studied most extensively thus far, the factor IX and DMD contigs have grown to the size of a typical cytogenetic band, and the factor VIII contig has resolved a difficult 1.6 Mb region that included a number of partially homologous repeated regions.

The formatting of the YAC contigs with hybridization probes has, however, proven to be difficult in some regions. An example is the region near the factor VIII gene, where there are several copies of highly homologous sequences indistinguishable by hybridization probes. Formatting a 1.6 Mb YAC contig was only possible with the extensive use of carefully chosen hybridization and PCR-based probes in the region (Freije and Schlessinger 1992).

Another type of problem is exemplified by the color vision locus, which has turned out to be highly unstable in YACs. The color vision genes are arranged as a set of three large tandemly repeated genes, and no YACs have been recovered to date that contain the entire set intact. Similar instability has been noted in YACs containing alphoid repeats (Neil et al. 1990) or repeat units of ribosomal DNA (Labella and Schlessinger 1989). Recombination between highly homologous, large tandemly repeated units may explain the commonly occurring internal deletions in YACs.

Although this type of approach is successful, it has several problems (Schlessinger et al. 1991). Especially annoying is that the formatting of the resultant map is neither simple nor obvious. In particular, if the map is to be formatted with PCR products of known sequence (STSs; Olson et al. 1989), they must be developed from the overlapping YACs after the map has been completed, a taxing and largely gratuitous activity once the ordered YACs are already assembled!

Evolution of mapping strategies

Hybridization-based methods in general suffer from the limitations mentioned above. Attempts to improve the efficiency of screening have employed high-density arrays of clones (Lehrach et al. 1990). A significant problem, however, remains the dependency on the reliable storage and exchange between laboratories of biological materials in large numbers, e.g., a resolution of 100 kb requires the use of at least 1500 markers per a 150-Mb chromosome. These problems can be largely avoided with the use of STSs directly as formatting reagents (for a detailed discussion, see Olson et al. 1989). The utility of STSs has also led the United States genome program to adopt them as uniquely defined map markers. The question then remains, how can STSs be derived most efficiently for use as mapping reagents? At least three approaches have been proposed and tested to some extent:

1. STSs can be derived independently of YACs by a method assuring random distribution along a chromosome. A wide variety of methods are available for the isolation of STS candidate sequences, including plasmid or phage clones from flow-sorted chromosomes or somatic cell hybrids (Green et al. 1991b), microdissection clones (Ludecke et al. 1989; MacKinnon et al. 1990), and inter-*Alu* PCR using cell hybrids containing whole or fragmented human chromosomes (Nelson et al. 1989; Cole et al. 1991).

2. STSs can be derived from YACs by a random approach. Here, the most efficient formulations rely on the use of inter-*Alu* PCR to isolate one or more single-copy fragments (Cole et al. 1991). Since YACs have been shown to be representative of the entire human genome, the distribution of STSs obtained with this approach should be essentially random.

3. STSs can be derived from the ends of YAC inserts ("all-walking approach"). This method is dependent on an efficient method to recover and sequence YAC insert ends (J. Kere et al., in prep.), but it has a number of advantages over the previously mentioned approaches. The representation of STSs should be essentially random as above in point 2, but in addition, the localization of the end-STSs within the source YAC are absolutely defined. Chimeric YACs are identified at once, and the contigs obtained with end-STSs from large YACs tend to be larger on average than those obtained with internal STSs as discussed by Palazzolo et al. (1991).

Our current formulation (J. Kere et al., in prep.) for the mapping of the human X chromosome therefore relies on the use of YAC insert end-STSs as primary map formatting reagents. Further increase in efficiency is achieved by coupling end-STS production to screening. The principle is to use primarily X-chromosome-specific YACs that have not yet been identified by STSs in the previous rounds of screening. When larger than average YACs are selected for STS development, each round of screening yields a maximal amount of previously uncovered DNA. To achieve sufficient marker density, other STSs can be added to the assembled contigs without the need to screen the entire YAC library with a large number of densely spaced markers. Using YAC libraries with an average insert size of 300 kb as sources of STSs and targets for screening, we have estimated that 500–600 STSs are needed to format contigs over more than 90% of the X chromosome, yielding an initial spacing of about 300 kb between STSs (J. Kere et al., in prep.).

Obviously, laboratories are not constrained to a single approach, and choices of methods may be made on the basis of local expertise and available materials. A typical factor is the quality and nature of YAC libraries at the disposition of a local group. For example, if a targeted library of YACs specific for a particular chromosome is on hand, mapping is more flexible, with map assembly considerably facilitated by the biological prefractionation.

HOW CAN THE MAP DRIVE THE STRUCTURAL ANALYSIS OF THE HUMAN GENOME?

If the initial goal of mapping is to cover a region with a formatted set of overlapping YACs, the further goal is to pursue "vertical integration" and preferably to the highest level of resolution, i.e., to arrive at the sequence of a region. Current strategies include subcloning YACs into λ or cosmid clones, as well as fragmentation vectors that contain elements present in the YAC insert (e.g., *Alu*, L1). The latter approach inserts new YAC vector elements within the original insert at the common elements to truncate the original YAC, and in this way, a nested series of deletions can be produced (Campbell et al. 1991; Pavan et al. 1991). However, no studies have reported any attempt to develop an optimized approach with various techniques. The intellectual and experimental challenge remains considerable, and one can anticipate rapid changes at the interface of subcloning and sequencing. Detailed discussion is beyond the scope of this chapter, but if sequencing methods improve to permit accurate determination of 1500 base pairs per run, the method of choice for cloning and the optimal mix of random and directed sequencing strategies would be very different.

Vertically integrated analysis of the genome can function with DNA as such: All sequences are more or less equal. However, the value of the resultant map for the scientific community is dependent not only on coverage of a region, but also on the rate and extent to which interesting features—the sequences that are more equal than others—are identified. For example, many would want to localize genes, fragile sites, telomeres, rare repetitive sequences, and rare-cutter restriction sites. In general, such elements can either be placed on the map at any time after regional map assembly is sufficiently complete or be incorporated with probes or STSs during the mapping itself. The choice is dependent on relative priorities and on the intrinsic ease of the placement. This can be seen for the two classes of sequences that are especially important for the paradigm of disease gene searches: the genetic linkage probes that detect polymorphism in the genome, and the potential units of function, transcribed or expressed regions. However, a PCR test developed from a moderately repetitive probe may give complex products from YAC pools in screening experiments, and some genes are not available in the form of sufficiently unique probes. For most of these cases, placement may be easier once a map is assembled, since only YACs that span the region of interest will be studied in detail.

YAC MAP-DRIVEN FUNCTIONAL ANALYSIS AND ITS ADVANTAGES

The question to be asked here is, how can the YAC-based map drive searches for functional units? This question can perhaps be clearly answered by considering current alternatives. At present, functional units are usually studied on a gene-by-gene basis or on an mRNA-by-mRNA basis. The standard situation is the study of a gene corresponding to a protein of interest to a particular investigator. In an extreme variation on the gene-by-gene approach, several groups are systematically cataloging and sequencing cDNAs (or portions of cDNAs) from random collections (Adams et al. 1991) in an attempt to carry out a global analysis of the functional capacity of the genome.

Although such approaches certainly sustain a fair fraction of the fruitful biochemistry now being done, map-based approaches can have several advantages. This is quite clear if one looks at three major categories of projected gene hunts.

Looking for specific disease genes

Map-driven analysis is becoming the most straightforward way to search for most disease genes. A hint of its prospective power comes from the instances of "candidate genes," cases in which likely genes already placed in a region are near enough to a genetically placed disease gene to be tested for coincidence. Examples are included in McKusick's recent review (1991). As more genes are mapped in contiguous DNA, the candidate gene approach will become increasingly potent, but at present, such approaches are directly dependent on prior map assembly.

There are already several examples in which YACs have facilitated mapping of a particular region and in which the map helped in identifying the gene of interest. The fragile-X syndrome was resolved following the discovery that a single YAC tested by in situ hybridization could cover the entire fragile region in metaphase chromosomes from affected individuals (Heitz et al. 1991; Kremer et al. 1991; Oberle et al. 1991; Verkerk et al. 1991; Yu et al. 1991); a search for NF1 on chromosome 17 (Wallace et al. 1990; D. Marchuk et al., pers. comm.) at the same time turned up YACs from the NF2 region of chromosome 22.

New methods supplement previous approaches that focused on chromosomal alterations in favorable patient material, such as deletions and translocations that interrupt or change the regulation of a gene of interest. For example, work in the laboratory of Andrea Ballabio has thus produced a deletion map of 17 intervals across Xp22.3, constructed from analysis of 44 patients. This study has revealed contiguous gene syndromes in the area; the complementary study of YACs from this region led to the isolation of a candidate gene for Kallman's syndrome (Franco et al. 1991). In a variant of this approach, Christine Petit and her collaborators (Legouis et al. 1991) have directly sequenced DNA in a likely region (see below) and located the same putative gene by its sequence characteristics.

Looking for groups of genes

The projects here are more broadly focused and may include all of the genes expressed in a tissue or all of the genes in a family (e.g., the major histocompatibility complex [MHC], tyrosine kinases, zinc finger proteins, the RAS-like family, and tubulins). For such goals, there are a priori advantages to searches for genes across YAC contigs compared to the analyses of one cDNA after another. By placing emphasis on map position rather than on homology only (as tested by hybridization), one can derive more conclusive information with less effort. For example, in the successive analysis of random cDNAs, one must reduce the chance of finding frequent cDNA species repeatedly. This leads to repetitive hybridization analyses with each cDNA against others or to the use of "normalized" cDNA libraries (Patanjali et al. 1991). Instead, when one works with YAC contigs, almost every cDNA will have only one locus in genomic DNA. Here, the number of cDNA clones positive for a particular location provides biologically useful information, since it can indicate the relative abundance of each mRNA in the cells or tissue from which the cDNA library was made.

In addition, the identification of "new" cDNAs for sequencing is facilitated by map-based approaches. Analyses of one cDNA after another are stymied by the need to determine whether a cross-hybridizing cDNA is derived from a gene already sequenced or represents another member of the family. If one sequences them all, one takes on the expense of considerable resequencing of the same cDNA. This problem is even more severe for genes like L1CAM, which produce many mRNA variants by alternative splicing (Cunningham et al. 1987). In that case, even sequencing may not be enough to discriminate provenance from one or more than one gene. On the other hand, if one attempts to make "normalized" cDNA libraries, many cross-hybridizing members of a gene family may be thrown away during the normalization processes, which are insensitive to the site of origin of the clone.

In contrast, with map-driven analysis, cross-hybridization of cDNAs is simply a measure of relatedness, rather than probable uniqueness; each member of a gene family is defined early in the analysis by its own chromosome location, even if loci are very tightly linked in the genome. In the example of L1CAM, the determination of a single specific location in cloned DNA is an unequivocal indication that the various mRNA species arise by alternative expression from one locus.

The representation and coverage of many cDNA libraries are largely unknown, so that one has no way to ensure convergence of a study to total coverage of the group of genes sought, and no criterion to assess how far one has advanced toward convergence. The problem is compounded because outgrowth may lead to the loss of rare or poorly propagating cDNA species. Instead, with map-driven analyses, progressive study of different cDNA libraries should provide an increasing number of genes in a region, which could converge toward saturation of expressed content.

Similar considerations apply to gene families defined in a number of ways, e.g., those containing a particular motif and those clustered together in a particular region. Examples of such motif-defined genes are the zinc finger proteins and tyrosine kinases. PCRs at low stringency and with degenerate primers have become a fruitful approach to isolate such groups of genes (Kamb et al. 1989; Wilks 1989; Nishi et al. 1990; Partanen et al. 1990; Pellegrino and Berg 1991). Again, map locations and the availability of corresponding genomic DNA provide ways to sort out the members of the family. Initially, panels of cell hybrids, flow-sorted DNA, and in situ hybridization provide chromosomal localizations. Then, even when there are syntenic localizations or multiple signals (through cross-hybridization), YAC contigs and panels provide detailed information to specify and sort out genes.

As work proceeds, observed features of the genome are providing additional strength to the a priori advantages of map-based approaches. For example, in the case of the Wilms' tumor region on chromosome 11, YACs were recovered from the Center of Genetics library at our university that include the putative Wilms' tumor locus (Bonetta et al. 1990). Searches by a number of methods (including direct screening of YACs against libraries of cDNAs; see below) then turned up a whole family of kidney-specific genes, dubbed an "archipelago" (Bonetta et al. 1990). There is no obvious way that the sequencing of one cDNA after another could easily have found the clustering of a number of genes that are related in tissue expression, possibly related in evolutionary origin, and perhaps coregulated during embryonic tissue development and function.

Until more of the human genome initiative goals are accomplished, it will not be known how general the "clustering" of genes by function or tissue distribution or time of expression may be. There are, however, many additional cases already known; they range from HLA (Bronson et al. 1991) to cytokines (Wilson et al. 1990) to the profusion of growth factor/oncogenes on chromosome 5 (Bishop and Westbrook 1990). In fact, current methods could be used to investigate, for example, the degree to which clustering occurs for the genes that specify an organelle or the characteristic features of a highly differentiated tissue.

Among the extreme examples that illustrate the importance of mapping down to the details of colinearity of genes are the analyses of homeobox (*HOX*) genes. Remarkable findings in both *Drosophila* and mice have indicated that the genes in tightly linked loci act sequentially in time during development and that their position in the cluster correlates with the anterior/posterior boundary of expression in the embryo gradients (Hunt and Krumlauf 1991). We do not know the basis for this extraordinary example of clustering, but it may well be conserved in corresponding mouse and human loci down to the level of individual members of the group of genes.

Looking for all genes

This is the largest-scale version of searching for genes, and subsumes the problems discussed above. A number of techniques have been used to find transcription units (genes) in genomic DNA (for a recent review, see Hochgeschwender and Brennan 1991); however, few combine simplicity and efficiency with the probability that a high fraction of the genes in the isolated genomic material are being detected. Once again, in approaches working with recursive sequencing of cDNAs, for example, there may be no use of cloned genomic DNA at all: A cDNA library can be prefractionated by subtractive hybridization to enrich for clones selectively expressed in a particular tissue or cell type, or cDNA libraries can be developed that tend to equalize the representation of various species (Patanjali et al. 1991). However, these searches discriminate poorly among closely related genes (requiring the separate study of multiple tissue-specific libraries) and require additional study to gain information about, for example, promoters, regulatory regions, and size of gene.

Comparable limitations apply to some map-based approaches as well. For example, CpG island searches in genomic or cloned DNA (Bird 1986; Lindsay and Bird 1987) provide an indication of possible gene content, but only a fraction of known genes have nearby CpG islands, and the study of genomic DNA is sensitive to tissue-specific methylation.

All of these techniques have been used to find transcribed regions of great interest, and this continues to be a very active area of study. They typically, however, provide no route to a complete census of genes in a region or to an accurate assessment of the degree of saturation achieved.

Can one instead devise *systematic* map-based functional analysis of genomic regions? The experimental challenge is to find robust ways to screen systematically for the constituent genes in a region. At present, methods are being developed that aim at the recovery of transcription units from preselected large regions of the genome and are in particular adaptable to contigs of overlapping YACs. Some of these methods are summarized in Table 3 and discussed below.

PRINCIPLES AND METHODS OF GENE SEARCHES AND UTILITY OF YACS

Gene searches based on partial or complete sequence information

Sequence information can be obtained by a variety of methods, including restriction enzyme digestion, oligonucleotide hybridization, and of course, sequencing. Each method can be adapted to gene searches. Perhaps the most traditional method is based on the observation that CpG dinucleotides are underrepresented in the genome, many of them are clustered and hypomethylated, and the clusters are often associated with 5' ends of genes. Thus, rare-cutter restriction enzymes can be conveniently used to identify regions potentially associated with genes (Bird 1986; Lindsay and Bird 1987). For CpG island identification, YACs provide distinct advantages. Large regions of genomic DNA can be conveniently covered and analyzed by restriction mapping of selected YACs. In addition, YACs are unmethylated, which allows for tissue-independent identification of rare-cutter islands, and further targeted subcloning of regions of interest is facilitated when cloned DNA is available in the form of YACs.

Oligonucleotide hybridization-based methods can be used to provide either high specificity for the desired sequence or redundancy for related sequences. Both properties can be used advantageously in filter-based hybridizations (Lehrach et al. 1990); however, PCR has provided

---

Table 3 Principles and methods of gene searches

Based on partial or complete sequence information
    Rare-cutter restriction enzyme maps and CpG island identification
    Identification of known regulatory or structural sequence motifs by PCR or hybridization
    Direct sequencing and identification of regulatory elements and open reading frames
Based on evolutionary conservation
    Zoo blots
    Cross-species hybridization and direct screening of cDNA libraries
Based on the study of expression or gene-specific functions in vivo
    Northern blots
    Direct screening of cDNA libraries by hybridization using phages, cosmids, or YACs as probes
    Hybridization-based enrichment and directed amplification
    Expression in hybrid cells containing fragments of foreign DNA
    Exon trapping and exon amplification: Functional identification of splicing signals
    Recombination-based assays between genomic DNA and cDNA

See text for details and examples of use.

strong alternatives. For example, members of a gene family can be searched using primers designed for the functionally most conserved regions of known members of the gene family. Such searches can be general, if genomic DNA is amplified, or targeted, if members of gene families are sought across a large cloned region. The amplified products can then be cloned and sorted. In both filter hybridizations and PCR, the sensitivity and specificity of assays can be manipulated by varying stringency.

In principle, oligonucleotides used in such experiments could include general signals such as transcriptional control elements (Locker and Buzard 1990). Here, with YACs as PCR templates, even moderately repetitive or relatively unspecific primers may provide considerable simplification of the analysis of the results, and even relatively poorly amplified signals may be adequate because the template is of low complexity and the background is correspondingly lower. The utility of this type of analysis should be especially evident in targeted searches across large YAC contigs.

Sequencing is always done in order to define a new gene, but definitive sequence information from long stretches of genomic DNA can also detect a new gene. Genes can be searched using computer algorithms to look for open reading frames, known signal sequences, or other general features of genes. The predictions for potential transcription units are then extended to experimental approaches to verify the presence of genes. The software to infer possible genes in raw sequence is progressively improving, but of course the large-scale application of sequencing to gene prediction is dependent on formidable increases in the capacity to generate long tracts of sequenced DNA, to the level where, for example, there would be little reason to try to sequence only expressed units of DNA rather than generating the entire sequence of cloned regions. Meanwhile, there are already hints of how such approaches might work. For example, in a study of sequences in λ and cosmid clones mapped across the factor VIII gene, another gene was discovered in intron 22 (Levinson et al. 1990). Similarly, analysis of 21 kb of the G6PD gene sequence has suggested the existence of another intronic transcript (Chen et al. 1991). Finally, in the case of Kallman's syndrome, Petit and colleagues sequenced portions of localized YAC DNA and inferred the location and identity of portions of the candidate gene (Legouis et al. 1991).

One way to concentrate on the analysis of portions of the genome that are highly enriched for genes has been pioneered by Bernardi's group. These authors showed that the genome is organized in regions of characteristic GC content that extend over hundreds of kilobases or more, and furthermore, that 3% of the total genome is very high in GC content and is also enriched on the order of tenfold in gene content (see, e.g., Aissani and Bernardi 1991). In effect, CpG islands are frequently diagnostic of gene positions because they tend to occur in the high-GC DNA fraction. In fact, recent studies of "compositional mapping" have shown directly that the highest gene concentrations in the genome are usually in "T-bands" (telomeric regions) of metaphase chromosomes (Saccone et al. 1992).

Furthermore, in an extensive study of Xq26-qter DNA cloned into overlapping YACs, it has been shown that high-GC regions can easily be recovered, even in cases where cytogenetic analysis may not have the resolving power to define a T band. For example, discontinuities in GC content are observed at or near the borders of cytogenetic bands; in Xq28, GC content rises progressively to reach 52% in a zone 2–4 Mb from the telomere (G. Pilia et al., in prep.). This zone is highly enriched for CpG islands and genes, including putative disease genes, to a level comparable to that observed in the nematode *C. elegans*. These characteristics make it a high priority region for extensive sequence analysis. In other words, it is already clear that about 300 Mb of the genome contains about 28% of the genes and that the corresponding portions of the genome are identifiable and recoverable by their specific mapping characteristics. Thus, information about sequence composition and overlapping clone maps can begin to drive the functional analysis of important regions of the genome.

Gene searches based on evolutionary conservation

Zoo blots or genomic DNA blots from a number of species have been used to search for genes in a large number of studies. The underlying principle is that many coding sequences serving important functions are well conserved during evolution. Conversely, conservation of a specific segment across distantly related species suggests functional significance. Again, zoo blots are only suggestive of the presence of a gene, and other techniques are needed to identify and confirm the gene. Generally, this method is applicable to phage or cosmid probes.

Two potentially powerful approaches combine the study of evolutionary conservation with direct screening of cDNA libraries or PCR amplification of DNA from one species using primers specific for other species. A recurrent problem in direct cDNA library screenings has been the high background signals created by repetitive sequences that are transcribed in many genes. Suppression of repeat hybridization by unlabeled competitor DNA helps to increase the signal-to-noise ratio, but this causes other problems. The problem can be essentially eliminated by using, for example, human YACs as probes against a library of mouse cDNAs. The notion that this would work came from the demonstration, building on early work of Murphy and Ruddle (1985), that even highly repetitive human DNA or rodent DNA, when used as a probe against YACs, essentially gives strong signals only from its species of origin (Wada et al. 1990). In contrast, general experience indicates that cDNAs or genes cross-hybridize comparatively well between human and mouse, even in exceptional cases like the X-inactivation-related Xist gene and its Xce mouse counterpart, which have diverged considerably during evolution (Borsani et al. 1991). In a number of control experiments (work in progress), this method has worked to find cDNAs with YACs. In fact, the studies show that human YACs can be used to detect corresponding regions in the mouse genome in zoo blots, and comparable tests of human YACs against mouse YACs can be envisaged.

Approaches based on sequencing can compare the sequences of a corresponding gene or region in various organisms to define the regions of interest and then test their importance by auxiliary assays including transfection studies. This approach, which again merges with long-range mapping strategies, is of interest because features of a sequence that are important for the function of a gene should be conserved during evolution. This may include not only the sequence of nucleotides, but also the spacing of translated, transcribed but not translated, and untranscribed regions. Such "pure" sequencing approaches have generally not been given serious consideration, because the amount of sequencing and the accuracy required is daunting. A rare partial example is that of the globin locus. In that case, duplications provoke gene conversion phenomena that are superimposed on other evolutionary changes and complicate any analysis, but some interesting inferences have been possible (Fitch et al. 1990).

Gene searches based on the study of expression or gene-specific functions in vivo

The study of expressed sequences is a straightforward way to identify genes, but it may often be laborious and insensitive. Hybridization of cloned DNA to Northern blots is limited to relatively short probes that must be relatively free of repetitive elements. Furthermore, only genes expressed in the tissues used as sources of mRNA can be detected. The limitation of probe length has been overcome in part in approaches where cDNA libraries are directly screened with genomic probes. With competitor DNA-based suppression of repetitive sequences, cDNA libraries have been screened with complex probes such as phages, cosmids, and YACs up to several hundred kilobases in length (Elvin et al. 1990). However, the sensitivity of detection is compromised with complex probes, and false positive signals, which require further sorting steps, further limit the usefulness of this approach.

More sophisticated applications of the hybridization-based comparisons between genomic and cDNA libraries aim at enriching and amplifying specific sequences derived from either the cDNA or genomic DNAs used in the experiment (Freije et al. 1991; Lovett et al. 1991; Parimoo et al. 1991). The hybridization-based enrichment is done on a solid support (e.g., a nylon filter or microbeads) and includes washes at desired stringency to regulate specificity of hybridization. The selected target DNAs are then recovered by PCR amplification for further study. These methods also require considerable follow-up to confirm potential genes and sort out false positive signals, but they combine conceptual simplicity, ease and speed of experimental work, and high sensitivity, and they have been primarily developed for gene searches over extended cloned regions such as YACs or YAC contigs. The number of tissues and the extent of the tracts of DNA that can be covered in a single experiment are still open to study. The development of these methods has now reached the stage where exhaustive searches for genes are being attempted on test cases of YAC contigs covering several megabases of genomic DNA.

Human genes can also be directly expressed in rodent hybrid cells containing part of a human chromosome. Some mRNAs contain human-specific repetitive elements that aid in the identification of human transcripts. If heterogeneous nuclear RNA (hnRNA) is used as a source for cDNA preparation, many more human genes can be directly identified by their content of unspliced human-specific repetitive elements. These techniques have been applied for the search of genes from specific human chromosomal regions (Neve et al. 1986; Liu et al. 1989; Corbo et al. 1990), but they might find new uses with YACs, or even better with YAC contigs, as genes encoded by YACs are expressed in mammalian cells (D'Urso et al. 1990; Gnirke et al. 1991; Huxley et al. 1991).

Elaborate techniques based on at least two different phenomena, splicing and recombination, have also been applied in vivo. These include exon trapping (Duyk et al. 1990), exon amplification (Buckler et al. 1991), and a recombination-based assay (Stewart and Kurnit 1990). All are potentially useful in the analysis of extended genomic regions, but most published protocols are awkward and require considerable follow-up work to sort out true positive clones. It is also not clear how comprehensively these methods detect different kinds of genes. Thus far, no tests of these techniques with YACs or YAC contigs have been reported.

HOW CAN THE MAP DRIVE THE TRACING OF THE COURSE OF EVOLUTION?

We have discussed above the use of comparative evolutionary sequence analysis to find genes. The unity of biochemistry is thereby increasingly apparent, based on the conservation of genes and other sequences in the chromosomal DNA among species. Once again, this leads to two alternatives for the analysis of evolution: gene-by-gene or map-based. Once again, both approaches are valid. For individual genes, direct sequence comparison can, as already recalled, provide information about conserved elements, and it can therefore provide fine-scale mapping of the course of evolutionary change. However, it has already become increasingly clear from comparative mapping studies that the identity and order of genes are also conserved in large syntenic blocks of up to 30 Mb or more along chromosomes across hundreds of millions of years of phylogeny. As a result, it seems feasible to analyze both the conserved elements and the course of change in equivalent regions of the genome across much of phylogeny.

One such attempt has shown that STSs for genes and other conserved sequences can be derived from human DNA, and work equally well for primates as distantly related as Old World monkeys, and that many STSs derived from gene sequences will also function in more distant species such as mouse, rat, and chicken (Mazzarella et al. 1992). Thus, one can imagine the use of a limited number of common conserved STSs to compare the organization of genomes across much of phylogeny. In its extreme form, as technology continues to gain the requisite power, one can anticipate a comparison of genomes on a YAC-by-YAC basis. The addition and movement of genes and repetitive sequences would be detected by the preservation of the evolutionary record in genomes.

CONCLUSION

The distinction between genetic and physical mapping will fade and the interdependence of the various ways to analyze the genome will become clear as the assembly of an ultimate single map begins. We have only begun to glimpse the ways in which the approaches treated above will interact. Thus, complete sequencing yields the map of DNA at high resolution, and tracts of sequence could in principle be assembled to produce a map of the genome, just as clones of overlapping DNA can be further analyzed to the level of sequence. Similarly, studies of conserved syntenically equivalent units can aid in the mapping of the human genome, and evolutionary comparisons of sequence can further the analysis of genes and control elements. In turn, the mapping of genes based on functional studies merges with genetic and comparative studies to provide ways to determine the course of evolution.

We infer that overlapping YACs provide a means to permit the conjoint progress of mapping and analysis of genomes. Optimism about the prospects for a unified map seems warranted, as the technology is already adequate to sustain genome projects and will continue to improve.

Acknowledgments

Many colleagues, including Maynard Olson, Phil Green, and Michele D'Urso, as well as our co-workers, have contributed to formulations presented here. The laboratory's work has been supported by National Institutes of Health grant HG-00201.

References

Adams, M.D., J.M. Kelley, J.D. Gocayne, M. Dubnick, M.H. Polymeropoulos, H. Xiao, C.R. Merril, A. Wu, B. Olde, R.F. Moreno, A.R. Kerlavage, W.R. McCombie, and J.C. Venter. 1991. Complementary DNA sequencing: Expressed sequence tags and human genome project. *Science* **252**: 1651.

Aissani, B. and G. Bernardi. 1991. CpG islands, genes, and isochores in the genomes of vertebrates. *Gene* **106**: 185.

Anand, R., J.H. Riley, R. Butler, J.C. Smith, and A.F. Markham. 1990. A 3.5 genome equivalent multi access YAC library: Construction, characterisation, screening and storage. *Nucleic Acids Res.* **18**: 1951.

Ballabio, A, B. Bardoni, R. Carrozzo, G. Andria, D. Bick, L. Campbell, B. Hamel, M.A. Ferguson-Smith, G. Gimelli, M. Fraccaro, R. Maraschio, O. Zuffardi, S. Guioli, and G. Camerino. 1989. Contiguous gene syndromes due to deletions in the distal short arm of the human X chromosome. *Proc. Natl. Acad. Sci.* **86**: 10001.

Barlow, D.P. and H. Lehrach. 1987. Genetics by gel electrophoresis: The impact of pulsed field gel electrophoresis on mammalian genetics. *Trends Genet.* **2**: 167.

Bird, A.P. 1986. CpG-rich islands and the function of DNA methylation. *Nature* **321**: 209.

Bishop, D.T. and C. Westbrook. 1990. Report of the committee on the genetic constitution of chromosome 5. *Cytogenet. Cell Genet.* **55**: 111.

Bishop, J.O. 1974. The gene numbers game. *Cell* **2**: 81.

Bodmer, W.F. 1981. Gene clusters, genome organization, and complex phenotypes. When the sequence is known, what will it mean? *Am. J. Hum. Genet.* **33**: 664.

Bonetta, L., S.E. Kuehn, A. Huang, D.J. Law, L.M. Kalikin, M. Koi, A.E. Reeve, B.H. Brownstein, H. Yeger, B.R. Williams, and A.P. Feinberg. 1990. Wilms tumor locus on 11p13 defined by multiple CpG island-associated transcripts. *Science* **250**: 994

Borsani, G., R. Tonlorenzi, M.C. Simmler, L. Dandolo, D. Arnaud, V. Capra, M. Grompe, A. Pizzuti, D. Muzny, C. Lawrence et al. 1991. Characterization of a murine gene expressed from the inactive X chromosome. *Nature* **351**: 325.

Bronson, S.K., J. Pei, P. Taillon-Miller, M.J. Chorney, D.E. Geraghty, and D.D. Chaplin. 1991. Isolation and characterization of yeast artificial chromosome clones linking the HLA-B and HLA-C loci. *Proc. Natl. Acad. Sci.* **88**: 1676.

Buckler, A.J., D.D. Chang, S.L. Graw, J.D. Brook, D.A. Haber, P.A. Sharp, and D.E. Housman. 1991. Exon amplification: A strategy to isolate mammalian genes based on RNA splicing. *Proc. Natl. Acad. Sci.* **88**: 4005.

Burke, D.T., G.F. Carle, and M.V. Olson. 1987. Cloning of large segments of exogenous DNA into yeast by means of artificial chromosome vectors. *Science* **236**: 806.

Butler, R., D.J. Ogilvie, P. Elvin, J.H. Riley, R.S. Finniear, G. Slynn, J.E.N. Morten, A.F. Markham, and R. Anand. 1992. Walking, cloning, and mapping with yeast artificial chromosomes: A contig encompassing D21S13 and D21S16. *Genomics* **12**: 42.

Campbell, C., R. Gulati, A.K. Nandi, K. Floy, P. Hieter, and R.S. Kucherlapati. 1991. Generation of a nested series of interstitial deletions in yeast artificial chromosomes carrying human DNA. *Proc. Natl. Acad. Sci.* **88:** 5744.

Chen, E.Y., A. Cheng, A. Lee, W.J. Kuang, L. Hillier, P. Green, D. Schlessinger, A. Ciccodicola, and M. D'Urso. 1991. Sequence of human glucose-6-phosphate-dehydrogenase cloned in plasmids and a yeast artificial chromosome. *Genomics* **10:** 792.

Coffey, A.J., R.G. Roberts, E.D. Green, C.G. Cole, R. Butler, R. Anand, F. Giannelli, and D.R. Bentley. 1992. Construction of a 2.6 Mb contig in yeast artificial chromosomes spanning the human dystrophin gene using an STS-based approach. *Genomics* **12:** 474.

Cole, C.G., P.N. Goodfellow, M. Bobrow, and D.R. Bentley. 1991. Generation of novel sequence tagged sites (STSs) from discrete chromosomal regions using Alu-PCR. *Genomics* **10:** 816.

Corbo, L., J.A. Maley, D.L. Nelson, and C.T. Caskey. 1990. Direct cloning of human transcripts with HnRNA from hybrid cell lines. *Science* **249:** 652

Coulson, A., Y. Kozono, B. Lutterbach, R. Shownkeen, J. Sulston, and R. Waterston. 1991. YACs and the *C. elegans* genome. *Bioessays* **13:** 413.

Cox, D.R., M. Burmeister, E.R. Price, S. Kim, and R.M. Myers. 1990. Radiation hybrid mapping: A somatic cell genetic method for constructing high-resolution maps of mammalian chromosomes. *Science* **250:** 245.

Cunningham, B.A., J.J. Hemperly, B.A. Murray, E.A. Prediger, R. Brackenbury, and G.M. Edelman. 1987. Neural cell adhesion molecule: Structure, immunoglobulin like domains, cell surface modulation, and alternative RNA splicing. *Science* **236:** 799.

Den Dunnen, J.T, P.M. Grootscholten, J.G. Dauwerse, A.P. Walker, A.P. Monaco, R. Butler, R. Anand, A.J. Coffey, D.R. Bentley, H.Y. Steensma, and G.J.B. van Ommen. 1992. Reconstruction of the 2.4 Mb human DMD gene by homologous YAC recombination. *Hum. Mol. Genet.* **1:** 19.

Donis-Keller, H., P. Green, C. Helms, S. Cartinhour, B. Weiffenbach, K. Stephens, T.P. Keith, D.W. Bowden, D.R. Smith, E.S. Lander, D. Bostein, G. Akots, K.S. Rediker, T. Gravius, V.A. Brown, M.B. Rising, L. Parker, J.A. Powers, D.E. Watt, E.R. Kauffman, A. Bricker, P. Phipps, H. Muller-Kahle, T.R. Fulton, S. Nig et al. 1987. A genetic linkage map of the human genome. *Cell* **51:** 319.

D'Urso, M., I. Zucchi, A. Ciccodicola, G. Palmieri, F.E. Abidi, and D. Schlessinger. 1990. Human glucose-6-phosphate dehydrogenase gene carried on a yeast artificial chromosome encodes active enzyme in monkey cells. *Genomics* **7:** 531.

Duyk, G.M., S.W. Kim, R.M. Myers, and D.R. Cox. 1990. Exon trapping: A genetic screen to identify candidate transcribed sequences in cloned mammalian genomic DNA. *Proc. Natl. Acad. Sci.* **87:** 8995.

Eliceiri, B., T. Labella, Y. Hagino, A. Srivastava, D. Schlessinger, G. Pilia, G. Palmieri, and M. D'Urso. 1991. Stable integration and expression in mouse cells of yeast artificial chromosomes harboring human genes. *Proc. Natl. Acad. Sci.* **88:** 2179

Elvin, P., G. Slynn, D. Black, A. Graham, R. Butler, J. Riley, R. Anand, and A.F. Markham. 1990. Isolation of cDNA clones using yeast artificial chromosome probes. *Nucleic Acids Res.* **18:** 3913.

Farr, C., J. Fantes, P. Goodfellow, and H. Cooke. 1991. Functional introduction of human telomeres into mammalian cells. *Proc. Natl. Acad. Sci.* **88**: 7006.

Fitch, D.H., C. Mainone, M. Goodman, and J.L. Slightom. 1990. Molecular history of gene conversions in the primate fetal gamma-globin genes. Nucleotide sequences from the common gibbon, *Hylobates lar. J. Biol. Chem.* **265**: 781.

Franco, B., S. Guioli, A. Pragliola, B. Incerti, B. Bardoni, R. Tonlorenzi, R. Carrozzo, E. Maestrini, M. Pieretti, P. Taillon-Miller, C.J. Brown, H.F. Willard, C. Lawrence, M.G. Persico, G. Camerino, and A. Ballabio. 1991. A gene deleted in Kallmann's syndrome shares homology with neural cell adhesion and axonal path-finding molecules. *Nature* **353**: 529.

Freije, D. and D. Schlessinger. 1992. A 1.6 Mb contig of yeast artificial chromosomes around the human factor VIII gene reveals a locus homologous to DXYS64 as well as several duplicated sequences. *Am. J. Hum. Genet.* **52**: (in press).

Freije, D., J. Kere, and D. Schlessinger. 1991. Enrichment of gene-specific sequences from YAC DNA. *Cytogenet. Cell Genet.* **58**: 1920.

Geraghty, D.E., J. Pei, B. Lipsky, J.A. Hansen, P. Taillon-Miller, S.K. Bronson, and D.D. Chaplin. 1992. Cloning and physical mapping of the HLA class I region spanning the HLA-E to HLA-F interval using yeast artificial chromosomes. *Proc. Natl. Acad. Sci.* **89**: 2669.

Gnirke, A., T.S. Barnes, D. Patterson, D. Schild, T. Featherstone, and M.V. Olson. 1991. Cloning and *in vivo* expression of the human GART gene using yeast artificial chromosomes. *EMBO J.* **10**: 1629.

Green, E.D. and P. Green. 1991. Sequence-tagged site (STS) content mapping of human chromosomes: Theoretical considerations and early experiences. *PCR Methods Appl.* **1**: 77.

Green, E.D. and M.V. Olson. 1990. Chromosomal region of the cystic fibrosis gene in yeast artificial chromosomes: A model for human genome mapping. *Science* **250**: 94.

Green, E.D., H.C. Riethman, J.E. Dutchik, and M.V. Olson. 1991a. Detection and characterization of chimeric yeast artificial chromosome clones. *Genomics* **11**: 658.

Green, E.D., R.M. Mohr, J.R. Idol, M. Jones, J.M. Buckingham, L.L. Deaven, R.K. Moyzis, and M.V. Olson. 1991b. Systematic generation of sequence-tagged sites for physical mapping of human chromosomes: Application to the mapping of human chromosome 7 using yeast artificial chromosomes. *Genomics* **11**: 548.

Heitz, D., F. Rousseau, D. Devys, S. Saccone, H. Abderrahim, D. Le Paslier, D. Cohen, A. Vincent, D. Toniolo, G. Della Valle, S. Johnson, D. Schlessinger, I. Oberlé, and J.L. Mandel. 1991. Isolation of sequences that span the fragile X and identification of a fragile X-related CpG island. *Science* **251**: 1236.

Hirst, M.C., K. Rack, Y. Nakahori, A. Roche, M.V. Bell, G. Flynn, Z. Christadoulou, R.N. MacKinnon, M. Francis, A.J. Littler, R. Anand, A.-M. Poustka, H. Lehrach, D. Schlessinger, M. D'Urso, V.J. Buckle, and K.E. Davies. 1991. A YAC contig across the fragile X site defines the region of fragility. *Nucleic Acids Res.* **19**: 3283.

Hochgeschwender, U. and M.B. Brennan. 1991. Identifying genes within the genome: New ways for finding the needle in a haystack. *BioEssays* **13**: 139.

Hourcade, D., A.D. Garcia, T.W. Post, P. Taillon-Miller, V.H. Holers, L.M. Wagner, N.S. Bora, and J.P. Atkinson. 1991. Analysis of the human regulators of complement activation (RCA) gene cluster with yeast artificial chromosomes (YACs). *Genomics* 12: 289.

Hunt, P. and R. Krumlauf. 1991. Deciphering the Hox code: Clues to patterning branchial regions of the head. *Cell* 66: 1075.

Huxley, C., Y. Hagino, D. Schlessinger, and M.V. Olson. 1991. The human HPRT gene on a yeast artificial chromosome is functional when transferred to mouse cells by cell fusion. *Genomics* 9: 742.

Kamb, A., M. Weir, B. Rudy, H. Varmus, and C. Kenyon. 1989. Identification of genes from pattern formation, tyrosine kinase, and potassium channel families by DNA amplification. *Proc. Natl. Acad. Sci.* 86: 4372.

Kerem, B., J.M. Rommens, J.A. Buchanan, D. Markiewicz, T.K. Cox, A. Chakravarti, M. Buchwald, and L.C. Tsui. 1989. Identification of the cystic fibrosis gene: Genetic analysis. *Science* 245: 1073.

Klinger, H.P., ed. 1991. Human gene mapping II: Eleventh International Workshop on human gene mapping. *Cytogenet. Cell Genet.* 58: 1.

Kozono, H., S.K. Bronson, P. Taillon-Miller, M.K. Moorti, I. Jamry, and D.D. Chaplin. 1991. Molecular linkage of the HLA-DR, HL-A-DQ, and HLA-DO genes in yeast artificial chromosomes. *Genomics* 11: 577.

Kremer, E.J., S. Yu, M. Pritchard, R. Nagaraja, D. Heitz, M. Lynch, E. Baker, V.J. Hyland, R.D. Little, M. Wada, D. Toniolo, A. Vincent, F. Rousseau, D. Schlessinger, G.R. Sutherland, and R.I. Richards. 1991. Isolation of a human DNA sequence which spans the fragile X. *Am. J. Hum. Genet.* 49: 656.

Labella, T. and D. Schlessinger. 1989. Complete human rDNA repeat units isolated in yeast artificial chromosomes. *Genomics* 5: 752.

Lagerstrom, M., J. Parik, H. Malmgren, J. Stewart, U. Pettersson, and U. Landegren. 1991. Capture PCR: Efficient amplification of DNA fragments adjacent to a known sequence in human and YAC DNA. 1991. *PCR Methods Applic.* 1: 111.

Legouis, R., J.-P. Hardelin, J. Levilliers, J.-M. Claverie, S. Compain, V. Wunderle, P. Millasseau, D. Le Paslier, D. Cohen, D. Caterina, L. Bougueleret, H. Delemarre-Van de Waal, G. Lutfalla, J. Weissenbach, and C. Petit. 1991. The candidate gene for the X-linked Kallmann syndrome encodes a protein related to adhesion molecules. *Cell* 67: 423.

Lehrach, H., R. Drmananc, J. Hoheisel, Z. Larin, G. Lennon, A.P. Monaco, D. Nizetic, G. Zehetner, and A. Poustka. 1990. Hybridization fingerprinting in genome mapping and sequencing. In *Genome analysis* (ed. K.E. Davies and S.M. Tilghman), vol. 1, p. 39. Cold Spring Harbor Laboratory Press, Cold Spring Harbor, New York.

Levinson, B., S. Kenwrick, D. Lakich, G. Hammonds, Jr., and J.A. Gitschier. 1990. Transcribed gene in an intron of the human factor VIII gene. *Genomics* 7: 1.

Lichter, P., C. Chang Tang, K. Call, G. Hermanson, G. Evans, D. Housman, and D. Ward. 1990. High resolution mapping of human chromosome 11 by *in situ* hybridization with cosmid clones. *Science* 247: 64.

Lindsay, S. and A.P. Bird. 1987. Use of restriction enzymes to detect potential gene sequences in mammalian DNA. *Nature* 327: 336.

Little, R.D., G. Pilia, S. Johnson, I. Zucchi, M. D'Urso, and D. Schlessinger. 1992. Yeast artificial chromosomes spanning 8 Mb and 10-15 centiMorgans of human cytogenetic band Xq26. *Proc. Natl. Acad. Sci.* 88: 177.

Liu, P., R. Legerski, and M.J. Siciliano. 1989. Isolation of human transcribed sequences from human rodent somatic cell hybrids. *Science* 246: 813.

Locker, J. and G. Buzard. 1990. A dictionary of transcription control sequences. *DNA Sequence* 1: 2.

Lovett, M., J. Kere, and L.M. Hinton. 1991. Direct selection: A method for the isolation of cDNAs encoded by large genomic regions. *Proc. Natl. Acad. Sci.* 88: 9628.

Ludecke, H.-J., G. Senger, U. Claussen, and B. Horsthemke. 1989. Cloning defined regions of the human genome by microdissection of banded chromosomes and enzymatic amplification. *Nature* 338: 348.

MacKinnon, R.N., M.C. Hirst, M.V. Bell, J.E. Watson, U. Claussen, H.J. Ludecke, G. Senger, B. Horsthemke, and K.E. Davies. 1990. Microdissection of the fragile X region. *Am. J. Hum. Genet.* 47: 181.

Mazzarella R., V. Montanaro, J. Kere, R. Reinbold, A. Ciccodicola, M. D'Urso, and D. Schlessinger. 1992. Conserved sequence-tagged sites: A phylogenetic approach to genome mapping. *Proc. Natl. Acad. Sci.* 89: 3681.

McKusick, V.A. 1990. *Mendelian inheritance in man*, 9th edition. Johns Hopkins University Press, Baltimore.

———. 1991. Current trends in mapping human genes. *FASEB J.* 5: 12.

Montanaro, V., A. Casamassimi, M. D'Urso, J.Y. Yoon, W. Freije, D. Schlessinger, M. Muenke, R.L. Nussbaum, S. Saccone, S. Maugeri, A.M. Santoro, S. Motta, and G. Della Valle. 1991. *In situ* hybridization to cytogenetic bands of yeast artificial chromosomes covering 50% of human Xq24-Xq28 DNA. *Am. J. Hum. Genet.* 48: 183.

Murphy, P.D. and F.H. Ruddle. 1985. Isolation and regional mapping of random X sequences from distal human X chromosome. *Somat. Cell Mol. Genet.* 11: 433.

National Research Council. 1988. *Mapping and sequencing the human genome.* National Academy Press, Washington.

Neil, D.L., A. Villasante, R.B. Fisher, D. Vetrie, B. Cox, and C. Tyler-Smith. 1990. Structural instability of human tandemly repeated DNA sequences cloned in yeast artificial chromosome vectors. *Nucleic Acids Res.* 18: 1421.

Nelson, D.L., S.A. Ledbetter, L. Corbo, M.F. Victoria, R. Ramirez-Solis, T.D. Webster, D.H. Ledbetter, and C.T. Caskey. 1989. Alu polymerase chain reaction: A method for rapid isolation of human-specific sequences from complex DNA sources. *Proc. Natl. Acad. Sci.* 86: 6686.

Neve, R.L., G.D. Stewart, P. Newcomb, M.L. Van Keuren, D. Patterson, H.A. Drabkin, and D.M. Kurnit. 1986. Human chromosome 21-encoded cDNA clones. *Gene* 49: 361.

Nishi, M., S. Ohagi, and D.F. Steiner. 1990. Novel putative protein tyrosine phosphatases identified by the polymerase chain reaction. *FEBS Lett.* 271: 178.

Oberle, I., F. Rousseau, D. Heitz, C. Kretz, D. Devys, A. Hanauer, J. Boue, M.F. Bertheas, and J.L. Mandel. 1991. Instability of a 550-base pair DNA segment and abnormal methylation in fragile X syndrome. *Science* 252: 1097.

Olson, M.V., L. Hood, C. Cantor, and D. Botstein. 1989. A common language for physical mapping of the human genome. *Science* 245: 1434.

Palazzolo, M.J., S.A. Sawyer, C.H. Martin, D.A. Smoller, and D.L. Hartl. 1991. Optimized strategies for sequence-tagged-site selection in genome mapping. *Proc. Natl. Acad. Sci.* 88: 8034.

Palmieri, G, V. Capra, G. Romano, M. D'Urso, S. Johnson, D. Schlessinger, P. Morris, J. Hopwood, P. Di Natale, R. Gatti, and A. Ballabio. 1992. The iduronate sulfatase gene: Isolation of a 1.2 Mb YAC contig spanning the entire gene and identification of heterogeneous deletions in patients with Hunter syndrome. *Genomics* 12: 52.

Parimoo, S., S.R. Patanjali, H. Shukla, D.D. Chaplin, and S.M. Weissman. 1991. cDNA selection: Efficient PCR approach for the selection of cDNAs encoded in large chromosomal DNA fragments. *Proc. Natl. Acad. Sci.* 88: 9623.

Partanen, J., T.P. Makela, R. Alitalo, H. Lehvaslaiho, and K. Alitalo. 1990. Putative tyrosine kinases expressed in K-562 human leukemia cells. *Proc. Natl. Acad. Sci.* 87: 8913.

Patanjali, S.R., S. Parimoo, and S.M. Weissman. 1991. Construction of a uniform-abundance (normalized) cDNA library. *Proc. Natl. Acad. Sci.* 88: 1943.

Pavan, W.J., P. Hieter, D. Sears, A. Burkhoff, and R.H. Reeves. 1991. High-efficiency yeast artificial chromosome fragmentation vectors. *Gene* 106: 125.

Pellegrino, G.R. and J.M. Berg. 1991. Identification and characterization of "zinc-finger" domains by the polymerase chain reaction. *Proc. Natl. Acad. Sci.* 88: 671.

Pfeifer, G.P., S.D. Steigerwald, P.R. Mueller, B. Wold, and A.D. Riggs. 1989. Genomic sequencing and methylation analysis by ligation mediated PCR. *Science* 246: 810.

Riley, J., R. Butler, D. Ogilvie, R. Finniear, D. Jenner, S. Powell, R. Anand, J.C. Smith, and A.F. Markham. 1990. A novel, rapid method for the isolation of terminal sequences from yeast artificial chromosome (YAC) clones. *Nucleic Acids Res.* 18: 2887.

Riordan, J.R., J.M. Rommens, B. Kerem, N. Alon, R. Rozmahel, Z. Grzelczak, J. Zielenski, S. Lok, N. Plavsic, J.L. Chou, M. Drumm, M.C. Iannuzzi, F.S. Collins, and L.C. Tsui. 1989. Identification of the cystic fibrosis gene: Cloning and characterization of complementary DNA. *Science* 245: 1066.

Rommens, J.M., M.C. Iannuzzi, B. Kerem, M.L. Drumm, G. Melmer, M. Dean, R. Rozmahel, J.L. Cole, D. Kennedy, N. Hidaka, M. Zsiga, M. Buchwald, J.R. Riordan, L.C. Tsui, and F.S. Collins. 1989. Identification of the cystic fibrosis gene: Chromosome walking and jumping. *Science* 245: 1059.

Rosenthal, A. and D.S.C. Jones. 1990. Genomic walking and sequencing by oligo-cassette mediated polymerase chain reaction. *Nucleic Acids Res.* 18: 3095.

Saccone S., A. De Sario, G. Della Valle, and G. Bernardi. 1992. The highest gene concentrations in the human genome are in T-bands of metaphase chromosomes. *Proc. Natl. Acad. Sci.* (in press).

Schlessinger D. 1990. Yeast artificial chromosomes: Tools for mapping and analysis of complex genomes. *Trends Genet.* 6: 248.

Schlessinger, D., R.D. Little, D. Freije, F. Abidi, I. Zucchi, G. Porta, G. Pilia, R. Nagaraja, S.K. Johnson, J.Y. Yoon, A. Srivastava, J. Kere, G. Palmieri, A. Ciccodicola, V. Montanaro, G. Romano, A. Casamassimi, and M. D'Urso. 1991. Yeast artificial chromosome-based genome mapping: Some lessons from Xq24-q28. *Genomics* 11: 783.

Silverman, G.A., E.D. Green, R.L. Young, J.I. Jockel, P.H. Domer, and S.J. Korsmeyer. 1990. Meiotic recombination between yeast artificial chromosomes yields a single clone containing the entire BCL2 protoon-cogene. *Proc. Natl. Acad. Sci.* **87**: 9913.

Silverman, G.A., J.I. Jockel, P.H. Domer, R.M. Mohr, P. Taillon-Miller, and S.J. Korsmeyer. 1991. Yeast artificial chromosome cloning of a two-megabase-size contig within chromosomal band 18q21 establishes physical linkage between BCL2 and plasminogen activator inhibitor type-2. *Genomics* **9**: 219.

Stewart, G.D. and D.M. Kurnit. 1990. Recombination-based screening for genes on chromosome 21. *Am. J. Med. Genet.* (suppl.) **7**: 115.

Verkerk, A.J., M. Pieretti, J.S. Sutcliffe, Y.H. Fu, D.P. Kuhl, A. Pizzuti, O. Reiner, S. Richards, M.F. Victoria, F.P. Zhang, B.E. Eussen, G.B. van Ommen, L.A.J. Blonden, G.J. Riggins, J.L. Chastain, C.B. Kirst, H. Gersaard, C.T. Caskey, D.L. Nelson, B.A. Oostra, and S.T. Warren. 1991. Identification of a gene (FMR-1) containing a CGG repeat coincident with a breakpoint cluster region exhibiting length variation in fragile X syndrome. *Cell* **65**: 905.

Wada, M., R.D. Little, F. Abidi, G. Porta, T. Labella, T. Cooper, G. Della Valle, M. D'Urso, and D. Schlessinger. 1990. Human Xq24-Xq28: Approaches to mapping with yeast artificial chromosomes. *Am. J. Hum. Genet.* **46**: 95.

Wallace, M.R., D.A. Marchuk, L.B. Andersen, R. Letcher, H.M. Odeh, A.M. Saulino, J.W. Fountain, A. Brereton, J. Nicholson, A.L. Mitchell, B.H. Brownstein, and F.S. Collins. 1990. Type 1 neurofibromatosis gene: Identi-fication of a large transcript disrupted in three NF1 patients. *Science* **249**: 181.

Wilks, A.F. 1989. Two putative protein-tyrosine kinases identified by application of the polymerase chain reaction. *Proc. Natl. Acad. Sci.* **86**: 1603.

Wilson, S.D., P.R. Billings, P. D'Eustachio, R.E. Fournier, E. Geissler, P.A. Lal-ley, P.R. Burd, D.E. Housman, B.A. Taylor, and M.E. Dorf. 1990. Cluster-ing of cytokine genes on mouse chromosome 11. *J. Exp. Med.* **171**: 1301.

Yin, S., M. Pritchard, E. Kremer, M. Lynch, J. Nancarrow, E. Baker, K. Holman, J.C. Mulley, S.T. Warren, D. Schlessinger, G.R. Sutherland, and R.I. Richards. 1991. Fragile X genotype characterized by an unstable region of DNA. *Science* **252**: 1179.

Zucchi, I. and D. Schlessinger. 1992. Distribution of moderately repetitive se-quences pTR5 and LFl in Xq24-q28 human DNA and their use in assem-bling YAC contigs. *Genomics* **12**: 264.

Index